图解建筑专业词汇

（原著第二版）

[英]罗科萨娜·麦克唐纳 著

张 杰 译

中国建筑工业出版社

著作权合同登记图字：01-2010-5590号

图书在版编目(CIP)数据

图解建筑专业词汇（原著第二版）/（英）麦克唐纳著；张杰译．—北京：中国建筑工业出版社，2011.6
ISBN 978-7-112-13155-6

Ⅰ．①图… Ⅱ．①麦…②张… Ⅲ．①建筑学–词汇–图解 Ⅳ．①TU-61

中国版本图书馆CIP数据核字（2011）第063264号

Copyright © 2007, Flora Samuel
All rights reserved

This edition of Illustrated Building Pocket Book by Roxanna McDonald is published by arrangement with ELSEVIER LIMITED of The Boulevard, Langford Lane, Kidlington, OXford, OX5 1GB, UK
The moral right of the author has been asserted

Translation© 2011 China Architecture & Building Press

本书由英国Elsevier出版社授权翻译出版

责任编辑：程素荣
责任设计：董建平
责任校对：陈晶晶　姜小莲

图解建筑专业词汇
（原著第二版）

[英] 罗科萨娜·麦克唐纳　著
张　杰　译

*

中国建筑工业出版社出版、发行（北京西郊百万庄）
各地新华书店、建筑书店经销
北京嘉泰利德公司制版
北京建筑工业印刷厂印刷

*

开本：787×1092毫米　1/32　印张：7 7/8　字数：230千字
2011年7月第一版　2011年7月第一次印刷
定价：28.00元
ISBN 978-7-112-13155-6
（20562）

版权所有　翻印必究
如有印装质量问题，可寄本社退换
（邮政编码100037）

目 录

前言　x
序言　xiii
致谢　xv
作者简介　xvii

第1章　建筑概要

1. 设计　3

模数比例　3
尺寸坐标　4
图纸表达　5
体积与形状　6

2. 作图技巧　7

作图实践——尺度，网格线　7
作图实践——比例与表现　8
作图工具　9
作图器具　10
电脑作图　11

3. 建筑类型　16

古典庙宇　16
古典柱式　17
古典装饰与增添装饰　18
中世纪装饰　19
中世纪城堡　20
风格比较　21
木结构框架建筑　22
典型教区教堂　23
哥特式主教座堂　24
传统房屋　25
大规模建造住宅　26
居住建筑　27
乡村建筑　31

传统农场建筑　　32

4.构件　33

入口　33
外部/内部　36
传统窗户　37

第2章　控制

1.法律类　41

英国土地法　41
阳光与天然光　42

2.管理　43

计划控制　43
规划批准　44
施工管理　45
伦敦建筑控制　46

第3章　建设过程

1.财政方面　49

开发　49
开发筹措资金法　50
评估与成本控制　51

2.项目执行　52

项目设计　52
项目施工　53
建筑承包合同　54

第4章　建筑工地

概述　57

定线（放线）　57
精确测量　58
测量设备　59

非破坏性测量：方法1　　60
非破坏性测量：方法2　　61
施工设备——起重机　　62
施工设备——挖掘　　63
脚手架——专利　　64
脚手架——独立　　65
混凝土施工设备　　66
动力工具　　67
焊接　　68
工具　　69

第5章　建筑构成

1.基础　　73

条形基础　　73
阶梯式基础　　74
柱下连续基础　　75
桩基础　　76
挡土墙　　77
柱下型钢基础　　78
支撑与托换基础　　79

2.上部结构——外墙　　80

砖墙施工　　80
砌砖　　81
砖砌合　　82
砖连拱　　83
砖类型　　84
潮湿的影响　　85
石材　　86
石材立面　　87
石材墙体　　88
琢石　　89
围护材料　　90
幕墙围护材料　　91
GRC围护材料　　92

3. 上部结构——内墙　93

　　隔墙　93
　　室内　96

4. 上部结构——屋盖　97

施工：
　　斜屋顶施工　97
　　斜屋顶类型　98
　　传统木屋顶类型　99
　　桁架　100
　　螺栓和桁架连接件　102
　　斜半屋架　103
　　桁架椽条组装　104
　　木结构节点　106
　　交叉木节点　107
　　传统嵌接节点　108
　　木模塑　110
　　平屋顶　111

屋面覆盖层：
　　屋面覆盖层类型　112
　　石板瓦　113
　　平瓦　114
　　波形瓦　115
　　茅草屋顶　116
　　传统木屋面板　117

屋盖开洞和防风雨：
　　屋盖开洞　118
　　屋顶窗　119
　　天窗　120
　　烟囱　121
　　防风雨——铅皮　122
　　防风雨——挡雨板　123
　　避雷装置　124
　　雨水管　125

5.上部结构——楼梯　　126

楼梯类型　　126
传统木楼梯　　127
特殊楼梯和自动扶梯　　128
电梯　　129

6.上部结构——烟囱　　130

壁炉　　130
壁炉附件　　131

7.上部结构——楼盖　　132

木楼盖结构　　132
木楼盖——饰边开洞　　133
木楼盖镶边　　134
拼板与角接　　135
混凝土楼盖　　136
钢筋混凝土楼盖　　137
顶棚——木托梁　　138
吊顶　　139
设备空间　　140
高架地板　　141

8.上部结构——墙上开洞　　142

门　　142
传统窗户——格窗　　146
传统窗户——竖铰链窗　　147
建筑五金——把手与门锁　　148
建筑五金——螺栓和把柄　　149
建筑五金——锁　　150
建筑五金——铰链　　151

9.上部结构——附件　　152

金属附件——系材　　152
金属附件——搁栅锚件　　153

金属部件——拉展金属网　　154
金属部件——过梁　　155
钉子与螺钉　　156
螺栓和栓头　　157

10.装修　　158

抹灰泥作业工具　　158
抹灰篱笆墙　　159
粉饰灰泥　　160
装饰性抹灰泥作业　　161
内墙——干衬砌　　162
木装修　　163
镶板　　164
地毯　　165

11.玻璃窗　　166

彩色玻璃　　166
平板玻璃　　167
玻璃安装系统　　168

12.服务——排水与管道系统　　169

排水　　169
地下排水　　170
管道系统、给水排水系统　　171
卫生管道作业　　172
冲洗式坐式便桶与储水箱　　173
管道系统连接　　174
水加热　　175

13.服务——电力　　176

电力——供应与布线　　176
电力——附件　　177
电力——布线路线　　178
户外照明　　179
电力装置　　180
电力——灯泡　　181

14. 服务——加热　182

加热系统　182
中央暖气系统——热水　183
散热器　184
空调　185

15. 户外作业/景观　186

土方工程——路边石　186
灌溉与屏蔽　187
围栏　188
树木　189
土地平整、草皮铺设　190
植物养护、附属房屋　191

第6章　环境状况

1. 全球变暖与温室效应　195

全球变暖与温室效应　195

2. 可持续绿色建筑　196

可持续绿色建筑　196
有关生物与气候的设计　197
材料功效——废物管理　198
能量效能设计　199

3. 危险中的建筑：自然灾害　201

地震　201
极端天气：飓风，闪电　204
块体运动：生成，滑坡　207
洪水　212
火山　213

英汉专业词汇对照　215

参考文献　232

前言

自从这本原名为《图解建筑词汇表》(An Illustrated Building Glossary)的图书在6年前出版以来,设计与管理的优先通道已经加速转变,对受过广泛教育的、怀有跨学科责任意识的专业人士的需求逐渐被理解。本书已经扩展到的这些领域见证了这个变化。

当然,对于环境保护的忧虑已列入专家的日程近20年,但公众是通过对自然资源挥霍使用的现象,才最终接受了由我们自身造成的现状的严重性。现今通常认为,存在于具有50年寿命的建筑物中的能量与资源仅有约1/4~1/3在它们的建设过程中被使用。因此,整个建筑的生命周期变成了工程初期的一个关键考虑因素。为了获得一个可持续的建筑环境,需要有能力促使它处于平衡的状态,每一个专业人员必须有对其发展进程的理解。电脑已成为一种手段,不仅勾画出我们的想法而且可以协调设计进程,预测建筑的性能并对其监测。当设计逐渐直接由电脑模型变为巧妙的产品,电脑便是办公用品采购时的重点对象。

1966年,当我加入位于伦敦W1区,菲茨罗伊街8号的奥雅纳(Ove Arup)公司时,在奥雅纳工程顾问公司及其合作伙伴的关于悉尼歌剧院壳体的设计中,一个大约25英尺高的地下室外墙需要一小时接着一小时地琢磨一些数字;让人想起令人印象深刻的无声电影《大都会》(Metropolis)的场景。而我现在坐在笔记本电脑前,却具有极大的潜能。如果我足够聪明,我可以只用一小口就吞下整个问题!这种变革现在允许设计思路被多方共享与尝试,同时,我们可以在一个新的四维空间里来优化解决方案。

但《图解建筑专业词汇》在10年后需要做哪些更新?乐观主义与悲观主义在这里交锋,因为我们生活在政治家的掌控之中,他们可以将我们推向每一个人需求的相反的方向。也许我可去猜测关于未来版本的任何变化。

从以碳为基础的经济向可持续能源的加速转变可以加强我们对于可再生的需求:这种变化可以通过一些途径成为现实,然而戏剧化的是,由于价格暴跌，太阳能发电的使用将变得普及，这需要我们的设计作出巧妙的变化。随着建筑物寿命周期的影响渗入行业，不仅仅是建筑构件要选择能使维护代价最低的，而且，购房人群逐渐开始从可能的寿命周期中的运作成本到资本投资来判断一个潜在的购买品。最后，不幸的却是，十年之后，当硬币最终落下来，无休止的经济增长会最终导致我们的失败而不是我们的赎回；从触及建筑环境的每一方面与我们获得它的方式都将会有大量的改变。

回到本书的内容：作为一名从业人员与大学教师，我实际上意识到把年轻学生与从业人员区分对待。当然，学生的知识并不全面，但是学习是一个持续的过程，通过一种新的图解的方式，重温描述复杂事物的词汇，将可以不仅重新估计那些事物，并且可以发掘源于这些事物的构思。

在建筑施工的实施过程中，所需知识与技能的范围是广泛的，从学习本书的任一页看出，这都是显而易见的。与每一幅插图中的某个条目相关的单词，仅仅是冰山一角的一处标记。每一个精心挑选过的单词都具有潜力；它介绍了建筑流程要素的一个部分，一个方面。相应地，建筑只是定义一个社会运作地点的开始；一个建筑物获得象征性存在的位置。所以你手里的这本书会激起你很多的共鸣。

这本书可以改变它的题目，但它仍是一个词汇表并且是与众不同的一本。这里，令人提神的是，解释是可视化的，通过"可视段落"清晰与完备的讲解，词义被给出，上下文被阐述，而这些通常需要文字来完成。事实上，作者仅有的文字书写部分是她的 500 个字的前言：一项令人称赞的成绩。

本书提供了什么？结构与对它们的描述值得作为设计的一部分来研究。大量的材料被挖掘，通过整合来清晰表述一个信息的核心而避免过于简单化。每一幅插图都来自于同一双手画了数个小时才完成，更无须提及相关的几个星期的研究、吸收与评价。任何喜欢争论设计细节或者喜欢尝试用一个简单的线条图画来陈述解决方法的人，都太清楚在着手画图之前花在图形琢磨上的时间了。

这类书只有在它使某个词条与其学科领域联系起来，并且从它出发可以到达更深入的参考书目时才会发挥作用。作者的参考书目不多但全部是常用的,通过那些书,可以探索更多的渠道。

归纳起来,这本书的价值在于对于有经验的专业人员或技术工人,它具有非常好的关于建设流程的直观信息,并通过吸引人的方式表达出来。它包含了许多可供学生学习的内容;对于有经验人士,它包含了许多我们曾经知道,但却不得不很惭愧地承认已经被我们遗忘的知识。

理查德·弗里沃

奥雅纳合伙人公司　董事 1977~2001 年
巴斯大学建筑学　讲座教授 1991~2000 年
香港大学建筑学　讲座教授 2000~2005 年

序 言

本书的主旨并非提供全面详尽的建筑术语或者尝试广泛地讲解建筑技术。在这方面有许多专业的百科全书、词典及建筑手册在这个领域提供充足的信息。本书是一部主要通过使用视觉参考来作为词汇交流的工具。

建筑的创造就是一个不同职业、观点，甚至国籍，并拥有各种技术知识与目的的人们之间相互合作的一个复杂过程的结果。居于中心位置的建筑师，经常发现他们就是参与者之间的"传译员"，并采用图像作为最保险的媒介。

我们每个人使用的语言来自我们自己的个人经历，有时，同一个词对不同的人可以代表不同的意思，这取决于他们学习它时的环境。这同样适用于建筑术语。

图像另一方面可尽量减少歧义，许多时候，工地上的问题与争论是通过潦草的画在墙上的草图得以解决的！

单词表达了我们对于可触摸物体的认识，并可以被归类进如按字母排序的字典或列于百科全书的上下文中。同理可以应用于图像——它们可以按字母排序排列于词汇附表上，或者它们可以被置于相关的上下文中。

这是本书采用的最新的体系，尝试在它们最有可能被应用的环境中表达这些术语。形成建筑语言的主要建筑术语采用按建筑流程的逻辑顺序来排列。如果一个人不记得恰当的词或者想知道一个特定部位如何称呼时，那从相关部分的草图中找出它将非常简单。类似地，学习一个术语，把它形象地放在某个环境中比记住它的抽象定义要容易得多。同时，反过来，索引使对于一个给定的词汇找到它的相关语境成为可能。

图画只是简单的草图，重在描述清晰、准确而非全面。图表意在确定顺序与关系以及特殊的术语。

作为一本主要为学生和刚刚参与工程的建筑人员编辑的可视化一览表，本书也意在供建筑业的其他从业者交流使用。

本书的真谛，我希望是重复了下面我所引用的一个古老的引言的蕴意。而时至今日，这个引言仍和它被第一次写下时一样有效。

只有有用的知识，使得一个人比另一个人更有价值，特别是直接关系到他赖以生存的那部分知识，因此，如果这本著作对于改进知识有助益，（为了主要预期），并且并不冒犯 fage Workman，重新唤醒他记忆中的法则，并告知其他从不认识的人，那将是净化他们心灵的最好的结果。

THO. Langley

1741 年 3 月 25 日，伦敦

来自引言：

THE BUILDER's JEWEL:
OR, THE
YOUTH's INSTRUCTOR,
AND
WORKMAN's REMEMBRANCER.

EXPLAINING

SHORT and EASY RULES,

Made familiar to the meanest Capacity,

For DRAWING and WORKING.

By B. and T. LANGLEY

LONDON.

Printed for R. Ware, at the Bible and Sun in Amen Corner, near Pater-Noster Row. MDCCXLI.   [Price 4. 6d.]

致　谢

对以下在本书第一版准备过程中给予支持的人和机构致以谢意：

Rob Dark, Architect, UK
B. Goilav, Structural Engineer, France
Dan S. Hanganu, Architect, Montreal, Canada
Claude and Anca Lemaire, Architects, France
Biblioteque Centre Pompidou, Paris, France
The RIBA Library, London, UK
Veronique Thierry, Isabelle Mathieu, Monique Beranger, Architects, Paris, France
Beatrice Jubien, France
Special thanks to Jane Fawcett whose generous advice and personal example were an inspiration.

特别感谢为本书带来灵感的简·福西特的慷慨的建议和个人的实例。

进一步感谢对筹备本书给予帮助的以下人士：

"工作或玩"（WORK OR PLAY）组织 Dominic Hailey (CAD 主任) 对于电脑绘图部分的建议。位于伦敦的多学科组织，"工作或玩"（www.workorplay.org）专门致力于 CAD 数据管理系统，它为英国与欧洲的建筑业的建筑师提供咨询与培训。

Jason Dunn，荣誉理学博士，英国建筑工程师协会会员，皇家特许测量师学会会员，MAIBS——协助更新《伦敦的施工管理与建筑控制》章节。

最后，我对我的编辑 Alex Hollingsworth 对于本书的完成给予的帮助、建议与可信赖的支持，致以最诚挚的感谢。

作者简介

罗科萨娜·麦克唐纳是一位执业建筑师,她在英国、法国和东欧工作,在一个广泛的专业领域——从历史建筑的保护到基础设施由于自然或人为灾难毁坏的重建,以及建筑相关的环境问题提供咨询。

作者写的其他书:

《壁炉书》(The Fireplace Book)——Architectural 出版社,1984 年

《图解建筑词汇表》(Illustrated Building Glossarg)——Butterworth Heineman 出版社,1999 年

《自然与人为灾难的介绍以及它们对建筑的影响》(Introduction to Natural and Man Made Disasters and their effects on Building)——Architectural 出版社,2003 年

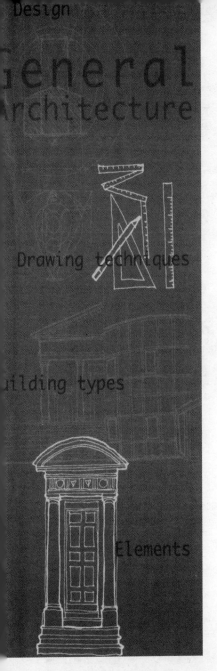

第1章 建筑概要

设计 —————— 3

作图技巧 —————— 7

建筑类型 —————— 16

构件 —————— 33

模数比例

1.设计

黄金数字
从古代起就被认为特别和谐
的尺寸比例：1.618
黄金断面 →
黄金数字的应用允许以一个
比例无限细分

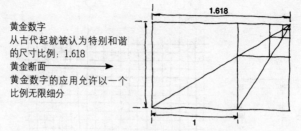

模数系统：勒·柯布西耶将黄金数字比例
应用于人体尺寸

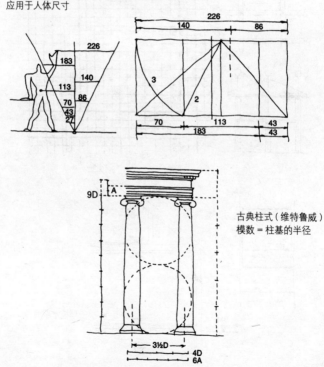

古典柱式（维特鲁威）
模数 = 柱基的半径

尺寸坐标

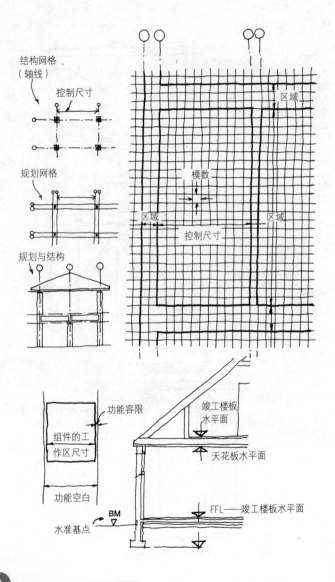

图纸表达

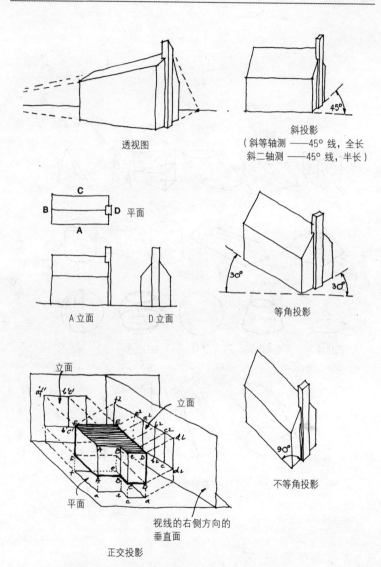

透视图

斜投影
（斜等轴测 ——45°线，全长
斜二轴测 ——45°线，半长）

平面

A立面　　D立面

等角投影

正交投影

不等角投影

体积与形状

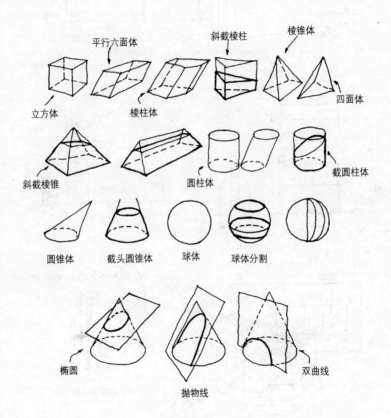

2.作图技巧

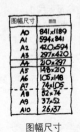

图幅尺寸

图幅尺寸	mm
A0	841×1189
A1	594×841
A2	420×594
A3	297×420
A4	210×297
A5	148×210
A6	105×148
A7	74×105
A8	52×74
A9	37×52
A10	26×37

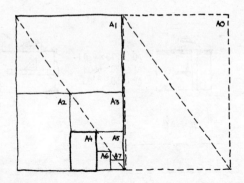

公制刻度标号

英制刻度标号

活动头手柄　　黄色1m记录　红色5m记录

测链　　　　　尺寸坐标

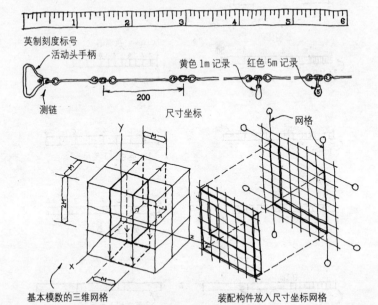

基本模数的三维网格　　　装配构件放入尺寸坐标网格

作图实践——比例与表现

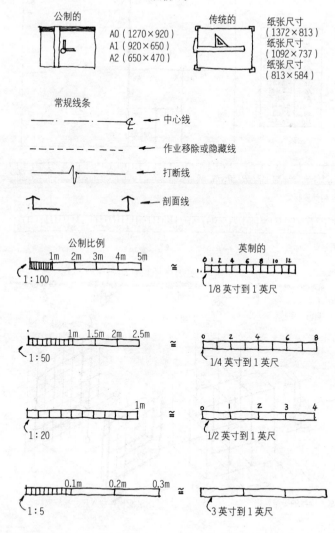

作图工具

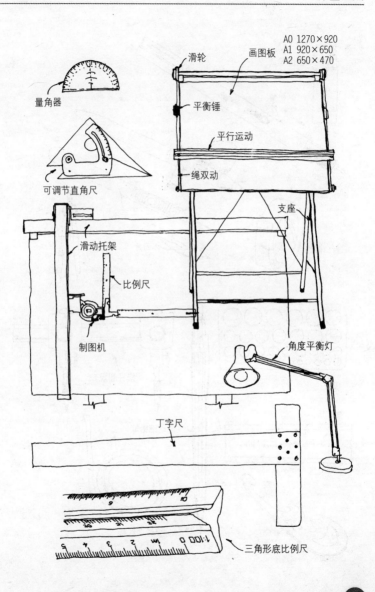

作图器具

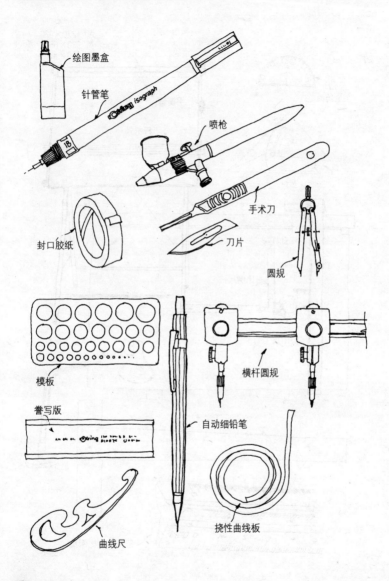

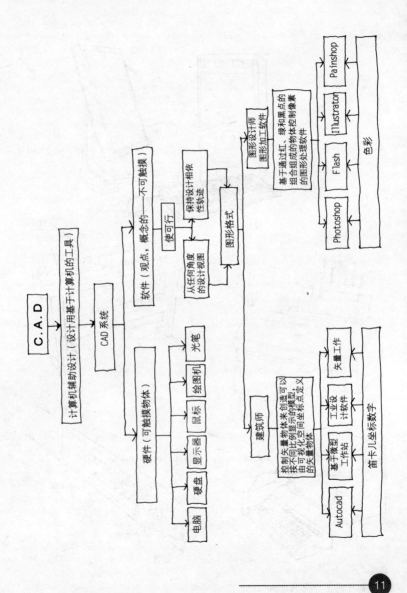

电脑作图

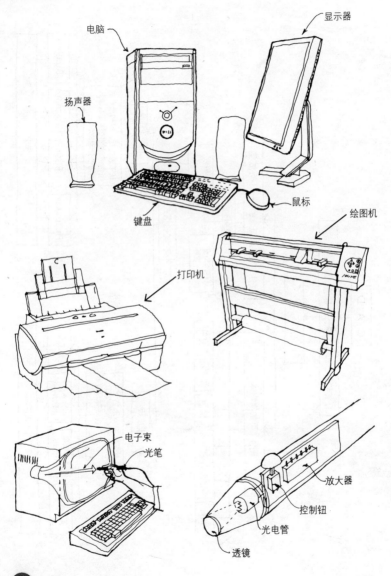

电脑作图

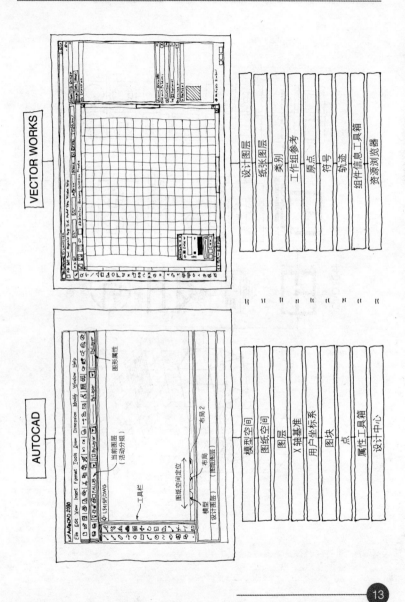

电脑作图

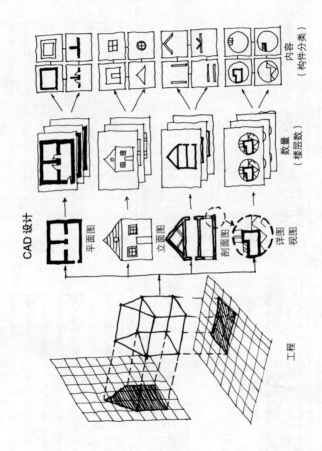

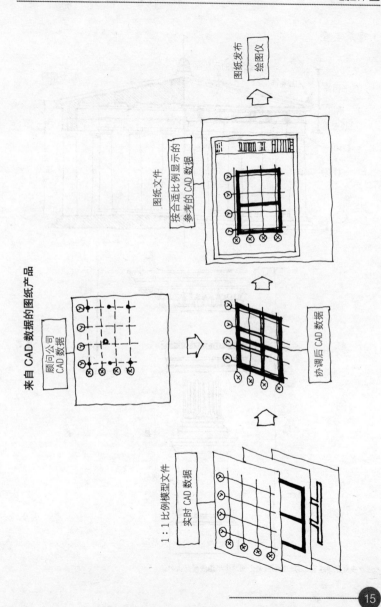

古典庙宇

3.建筑类型

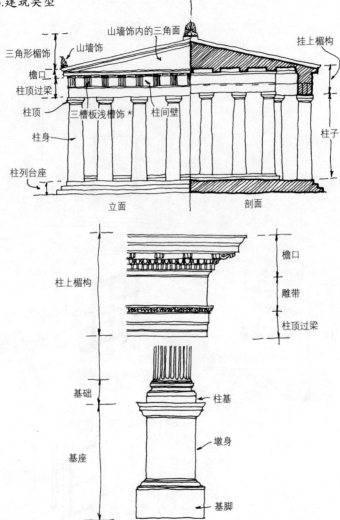

*注：多立克式建筑上的三竖面或两个和两个半边浅槽饰。

古典柱式

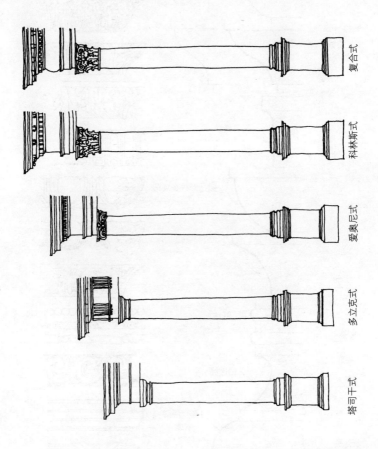

复合式
科林斯式
爱奥尼式
多立克式
塔司干式

古典装饰与增添装饰

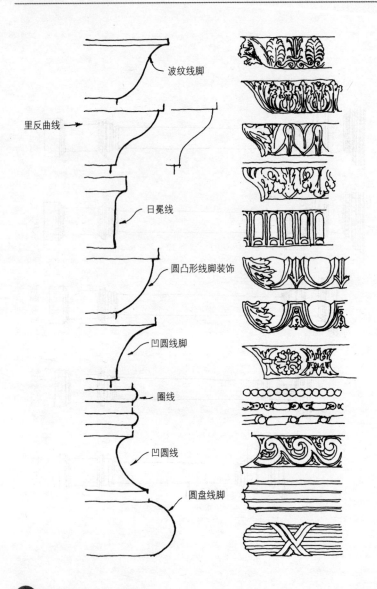

- 波纹线脚
- 里反曲线
- 日冕线
- 圆凸形线脚装饰
- 凹圆线脚
- 圈线
- 凹圆线
- 圆盘线脚

中世纪装饰

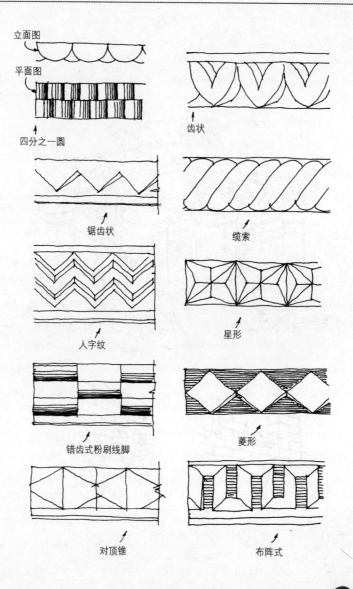

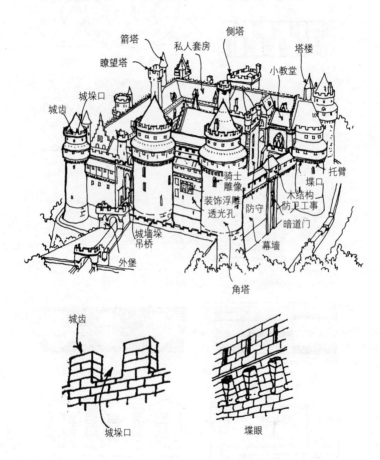

风格比较

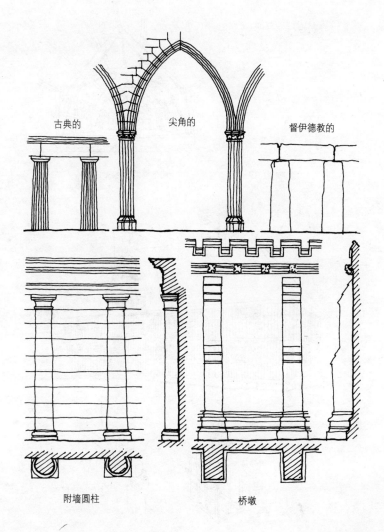

木结构框架建筑

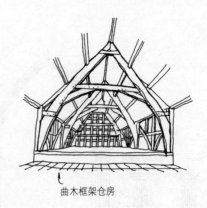

曲木框架仓房

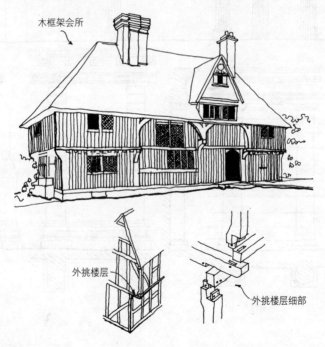

木框架会所

外挑楼层

外挑楼层细部

典型教区教堂

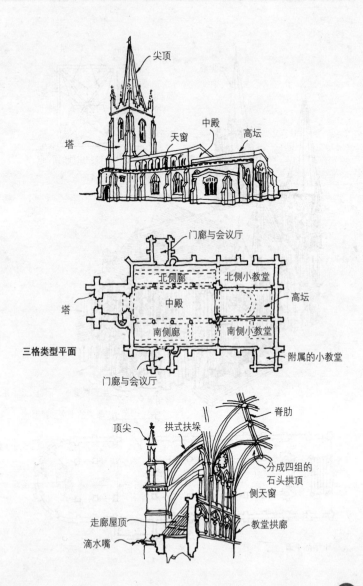

三格类型平面

哥特式主教座堂

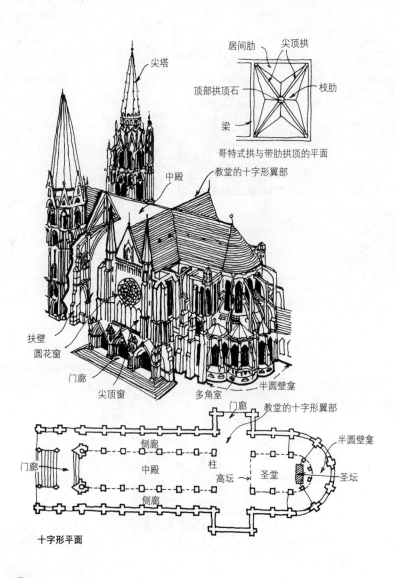

传统房屋

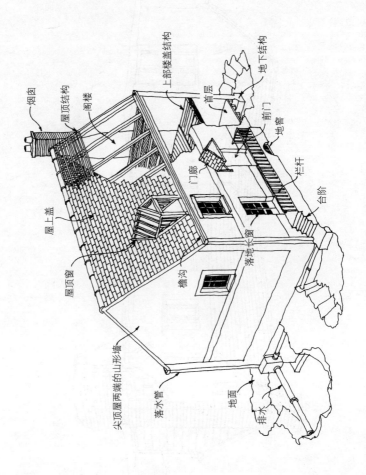

大规模建造住宅

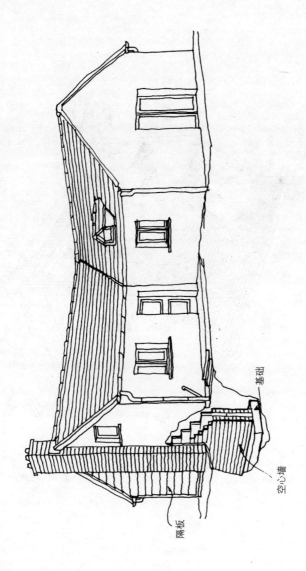

居住建筑

联立式

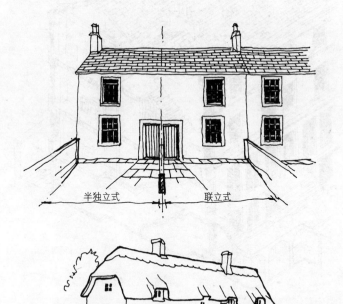

半独立式　　联立式

独立式小屋

居住建筑

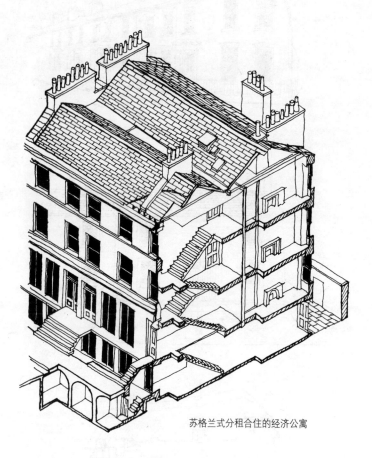

苏格兰式分租合住的经济公寓

居住建筑

富丽堂皇的住宅

文艺复兴时的城堡

中世纪的城堡

居住建筑

独立式住宅

平房

高层公寓楼

乡村建筑

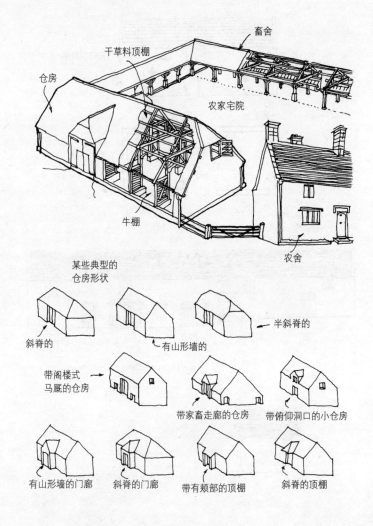

传统农场建筑

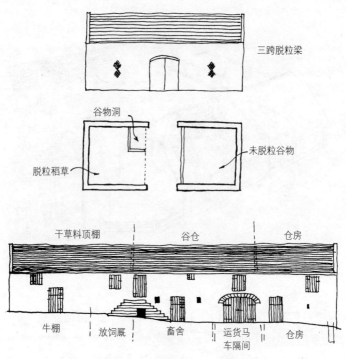

三跨脱粒梁

谷物洞

脱粒稻草

未脱粒谷物

干草料顶棚　谷仓　仓房

牛棚　放饲厩　畜舍　运货马车隔间　仓房

储藏室

前开式棚开间

入口

4.构件

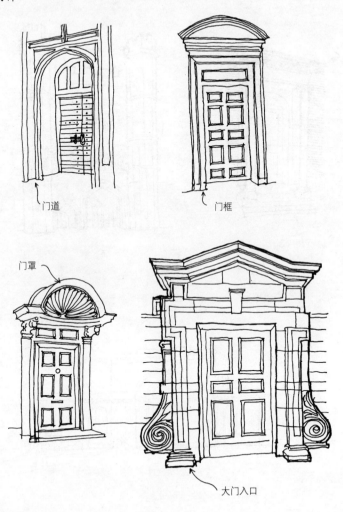

门道　　门框　　门罩　　大门入口

入口

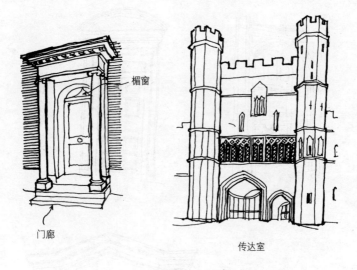

楣窗

门廊

传达室

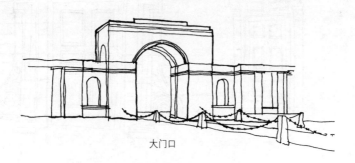

大门口

入口

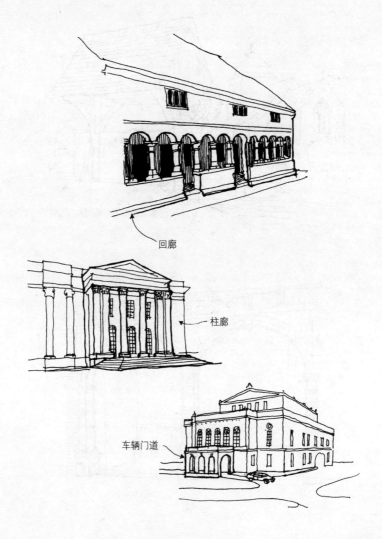

回廊

柱廊

车辆门道

外部/内部

教堂墓地中有顶的大门（停柩门）

凉廊

传统窗户

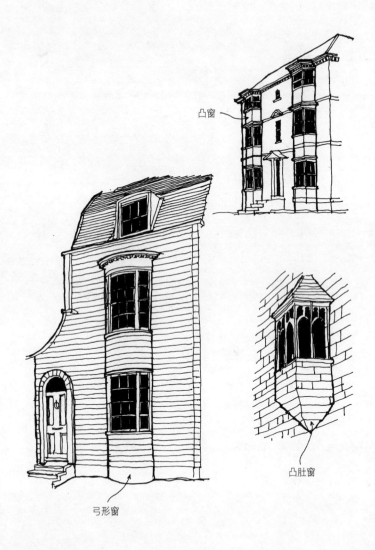

凸窗

凸肚窗

弓形窗

第2章 控制

法律类 —— 41

管理 —— 43

英国土地法

1.法律类

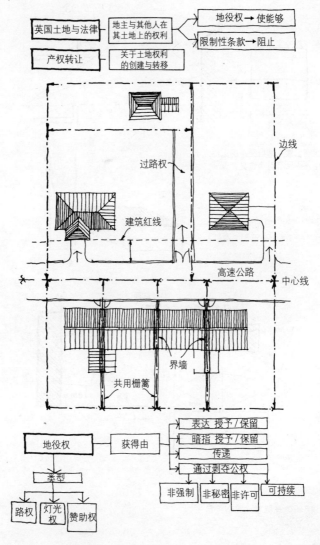

阳光与天然光

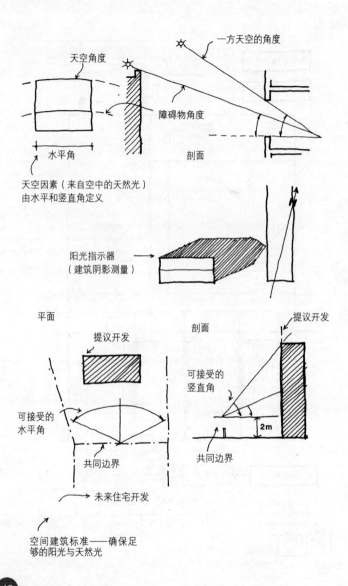

天空角度

水平角

天空因素（来自空中的天然光）由水平和竖直角定义

一方天空的角度

障碍物角度

剖面

阳光指示器（建筑阴影测量）

平面

提议开发

可接受的水平角

共同边界

未来住宅开发

剖面

提议开发

可接受的竖直角

2m

共同边界

空间建筑标准——确保足够的阳光与天然光

计划控制

2.管理

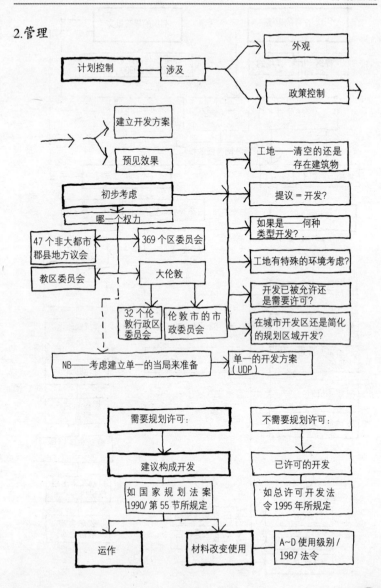

规划批准

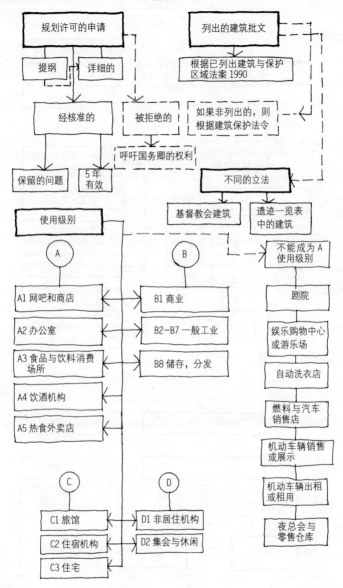

施工管理

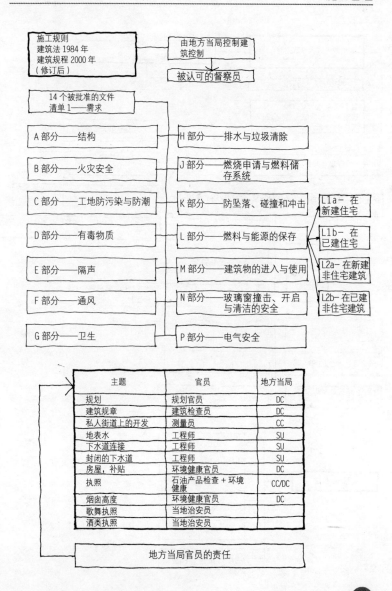

伦敦建筑控制

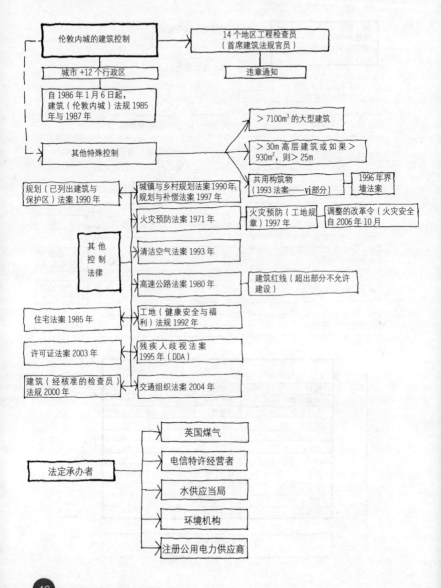

第3章 建设过程

财政方面 —— 49

项目执行 —— 52

开发

1.财政方面

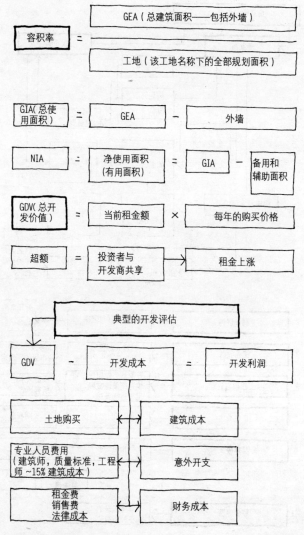

开发筹措资金法

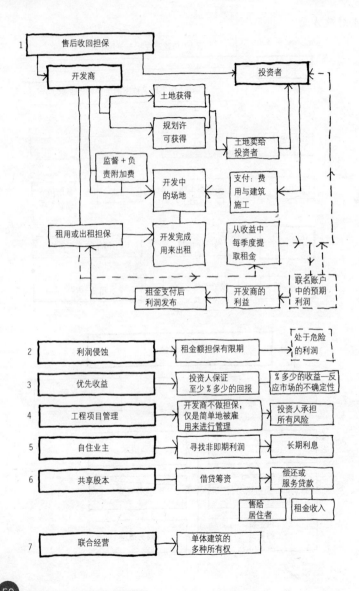

评估与成本控制

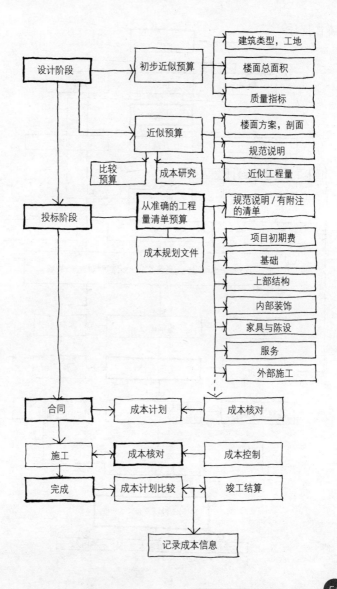

项目设计

2.项目执行

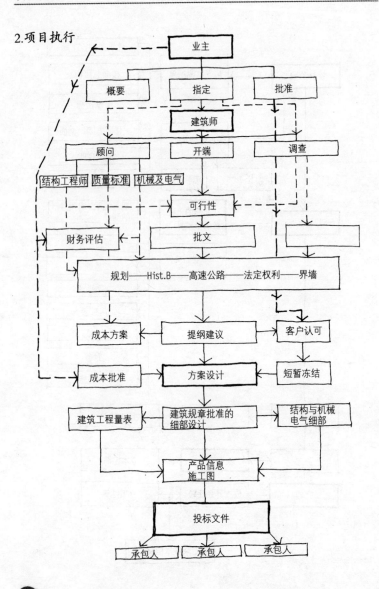

项目施工

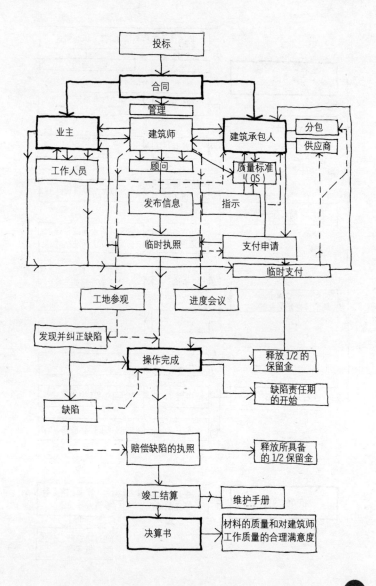

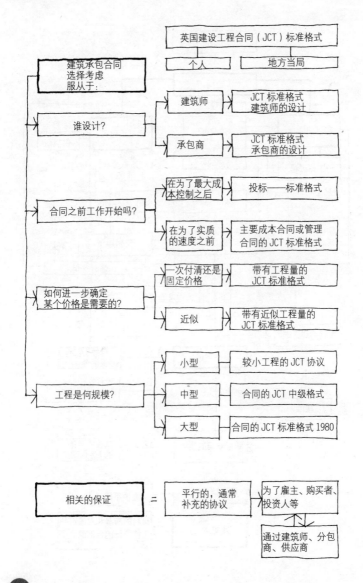

第4章 建筑工地

概述 57

定线（放线）

概述

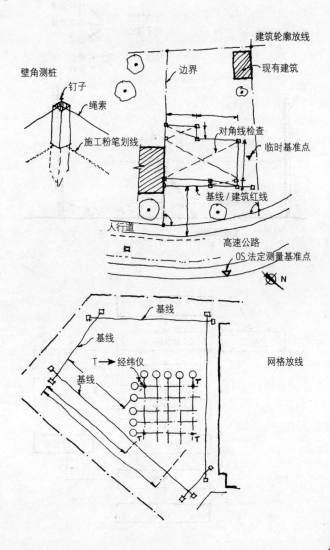

精确测量

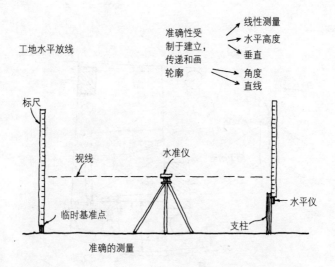

准确的测量

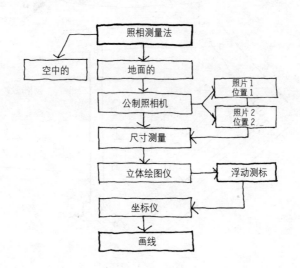

测量设备

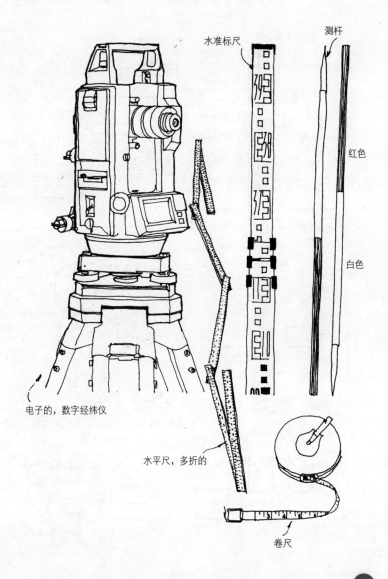

水准标尺

测杆

红色

白色

电子的，数字经纬仪

水平尺，多折的

卷尺

非破坏性测量：方法1

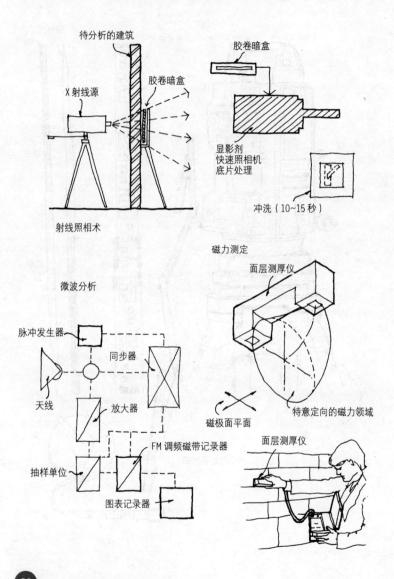

非破坏性测量：方法 2

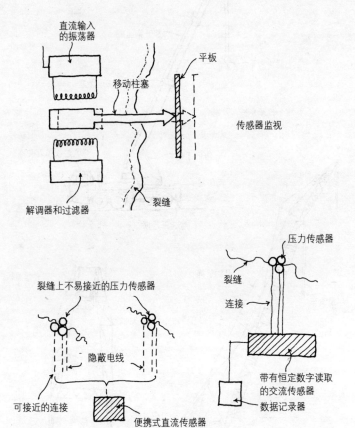

施工设备——起重机

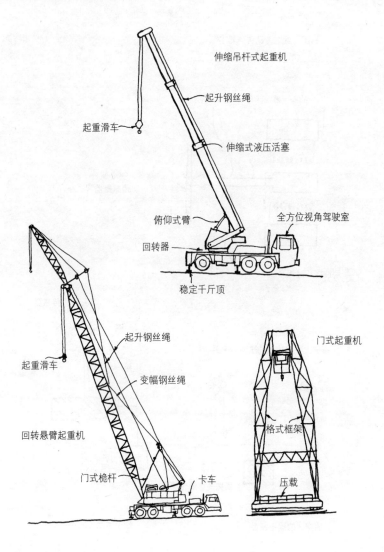

施工设备——挖掘

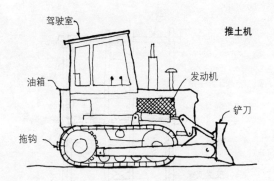

推土机

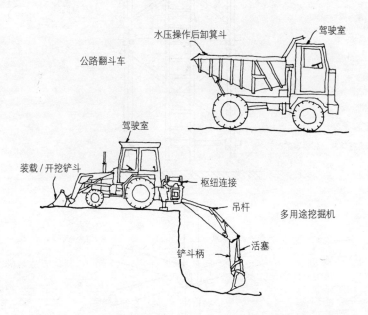

公路翻斗车

多用途挖掘机

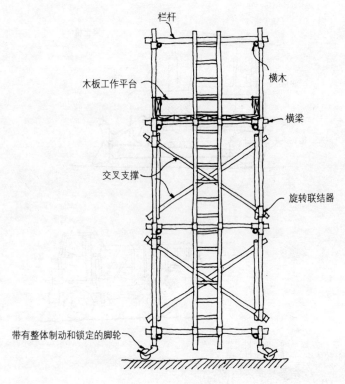

脚手架塔

栏杆
横木
木板工作平台
横梁
交叉支撑
旋转联结器
带有整体制动和锁定的脚轮

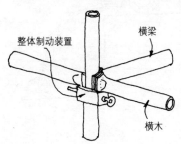

专利脚手架

整体制动装置
横梁
横木

脚手架——独立

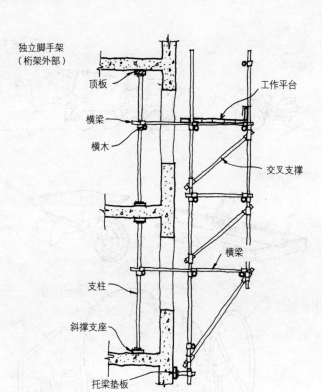

独立脚手架
（桁架外部）
顶板
横梁
横木
工作平台
交叉支撑
横梁
支柱
斜撑支座
托梁垫板

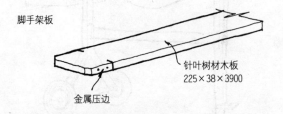

脚手架板
针叶树材木板
225×38×3900
金属压边

混凝土施工设备

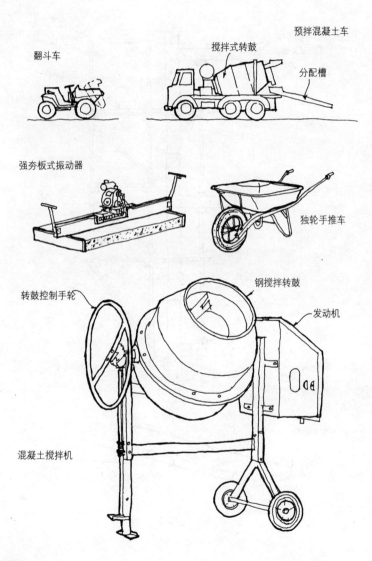

动力工具

发电机组

夯具

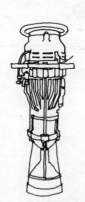
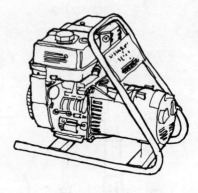

压缩机

风镐

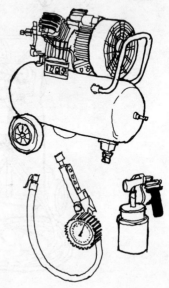

焊接

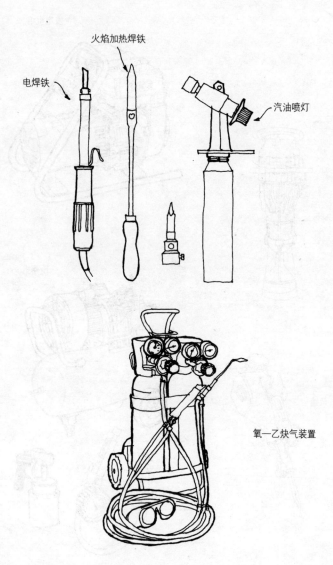

电焊铁
火焰加热焊铁
汽油喷灯
氧—乙炔气装置

工具

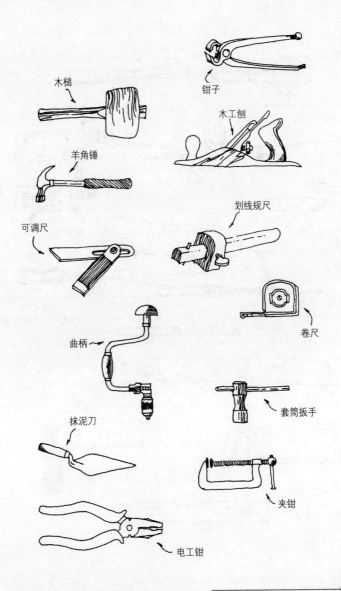

工具

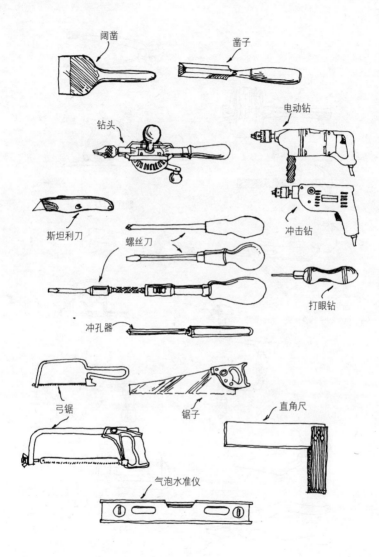

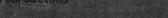

第5章 建筑构成

基础 —— 73

上部结构——外墙 —— 80

上部结构——内墙 —— 93

上部结构——屋盖 —— 97

上部结构——楼梯 —— 126

上部结构——烟囱 —— 130

上部结构——楼盖 —— 132

上部结构——墙上开洞 —— 142

上部结构——附件 —— 152

装修 —— 158

玻璃窗 —— 166

服务——排水与管道系统 —— 169

服务——电力 —— 176

服务——加热 —— 182

户外作业/景观 —— 186

条形基础

1. 基础

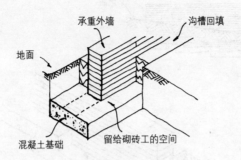

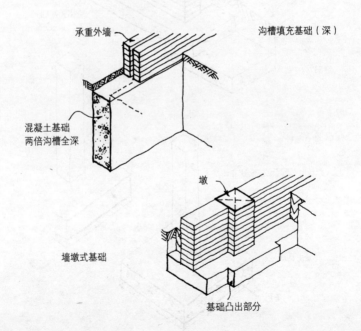

阶梯式基础

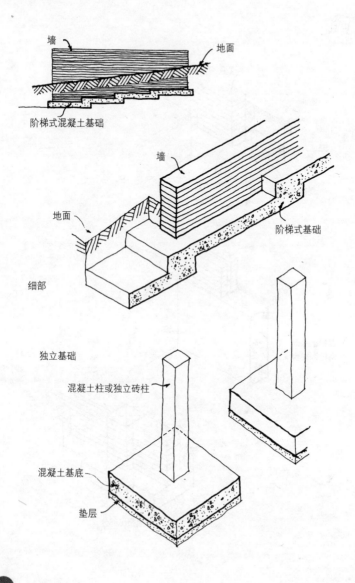

柱下连续基础

对于间隔排列很近的柱子

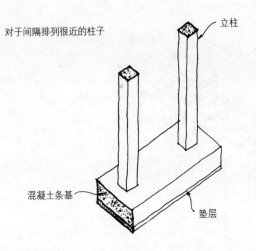

立柱

混凝土条基

垫层

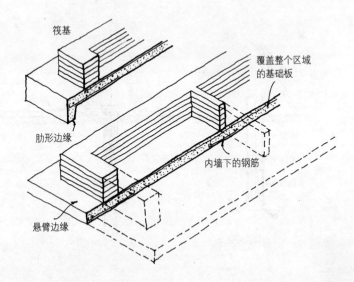

筏基

覆盖整个区域的基础板

肋形边缘

内墙下的钢筋

悬臂边缘

桩基础

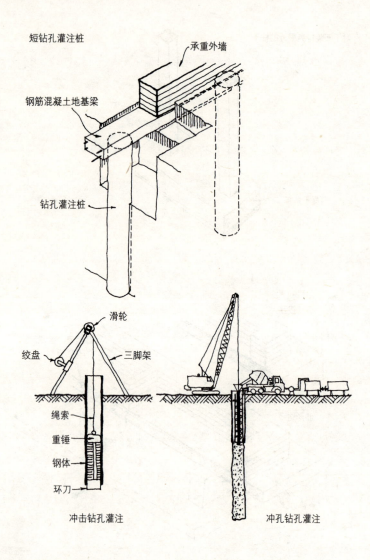

短钻孔灌注桩
承重外墙
钢筋混凝土地基梁
钻孔灌注桩

滑轮
绞盘
三脚架
绳索
重锤
钢体
环刀

冲击钻孔灌注

冲孔钻孔灌注

挡土墙

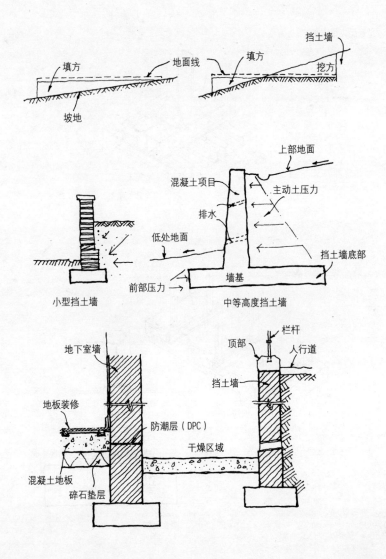

柱下型钢基础

型钢格排基础

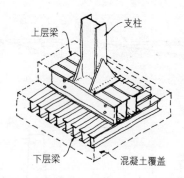

复合柱下型钢基础

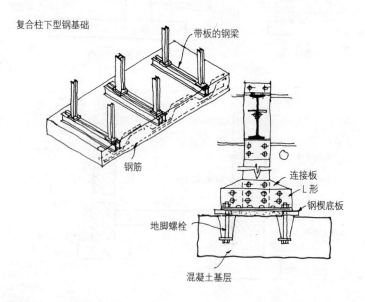

支撑与托换基础

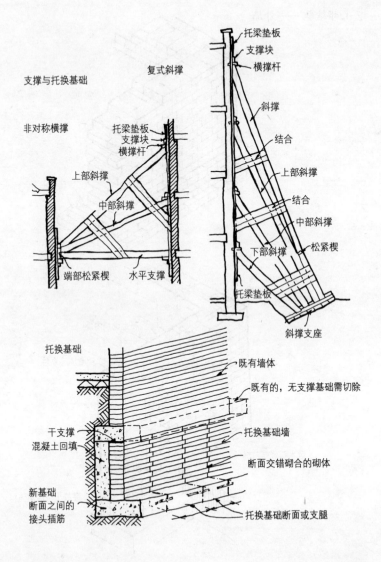

砖墙施工

2.上部结构——外墙

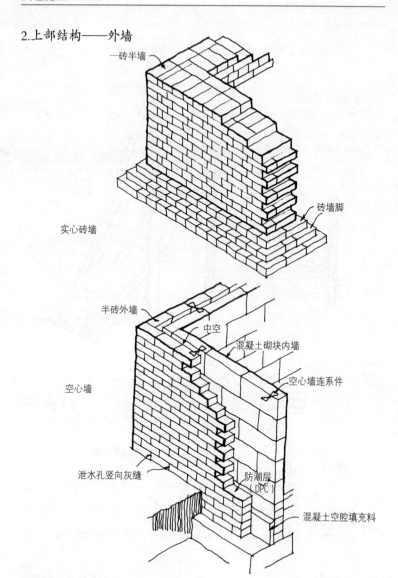

一砖半墙

砖墙脚

实心砖墙

半砖外墙

中空

混凝土砌块内墙

空心墙连系件

空心墙

泄水孔竖向灰缝

防潮层（DPC）

混凝土空腔填充料

砌砖

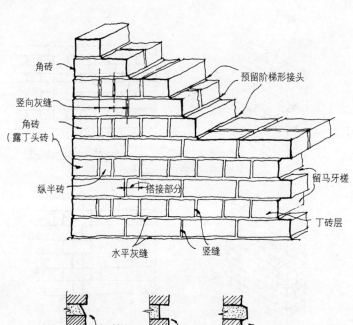

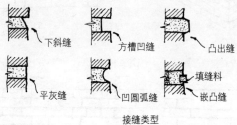

接缝类型

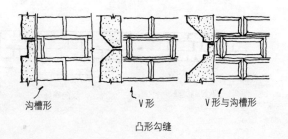

凸形勾缝

砖砌合

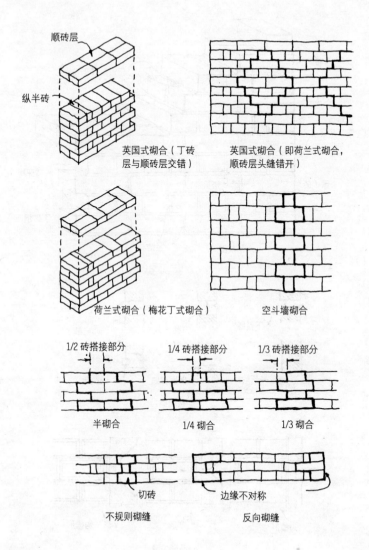

砖连拱

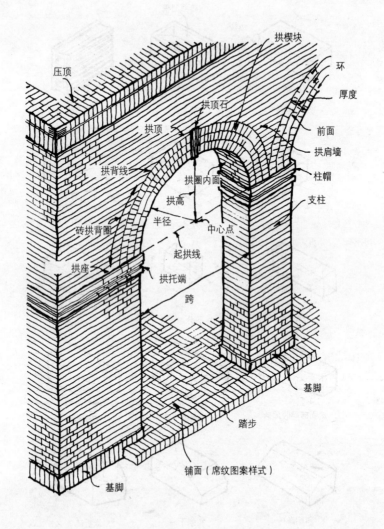

砖类型

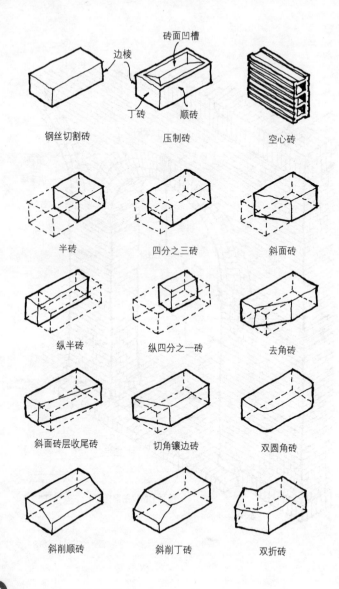

潮湿的影响

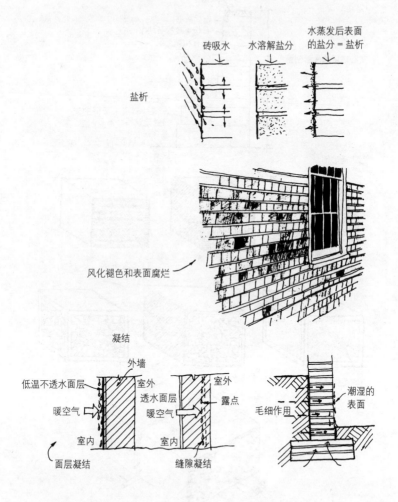

石材

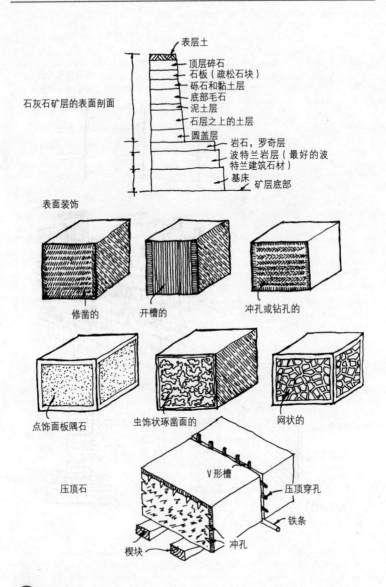

石材立面

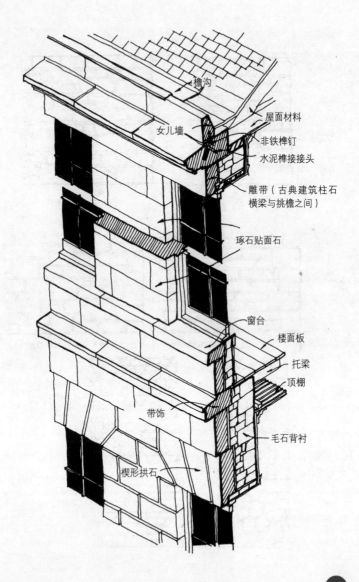

石材墙体

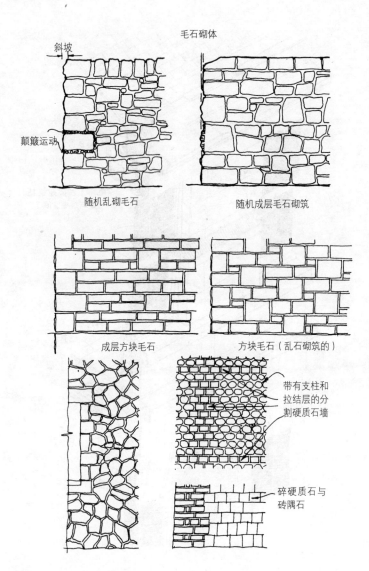

毛石砌体

斜坡

颠簸运动

随机乱砌毛石

随机成层毛石砌筑

成层方块毛石

方块毛石（乱石砌筑的）

带有支柱和拉结层的分割硬质石墙

碎硬质石与砖隅石

琢石

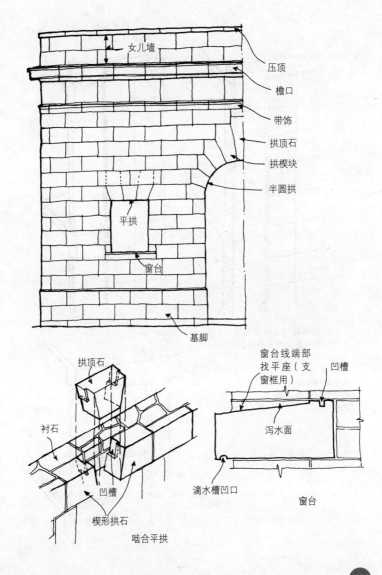

围护材料

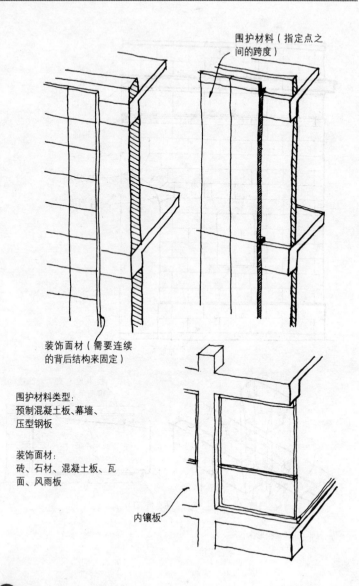

围护材料（指定点之间的跨度）

装饰面材（需要连续的背后结构来固定）

围护材料类型：
预制混凝土板、幕墙、压型钢板

装饰面材：
砖、石材、混凝土板、瓦面、风雨板

内镶板

幕墙围护材料

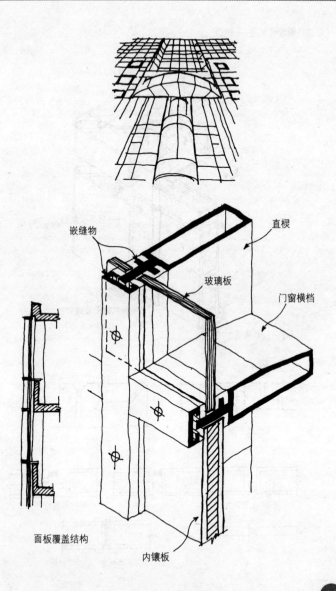

GRC 围护材料

（玻璃纤维增强水泥——GRC）

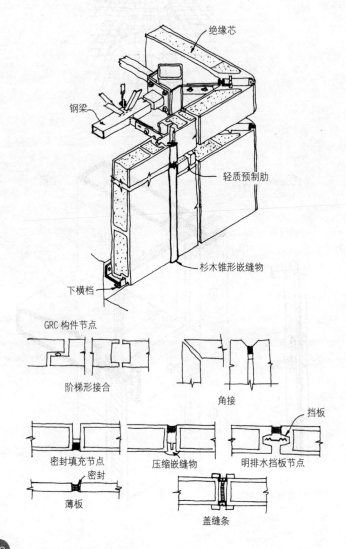

3. 上部结构——内墙

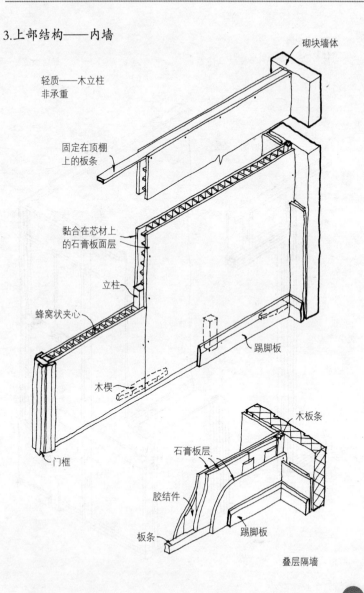

叠层隔墙

隔墙

内部——金属立柱非承重

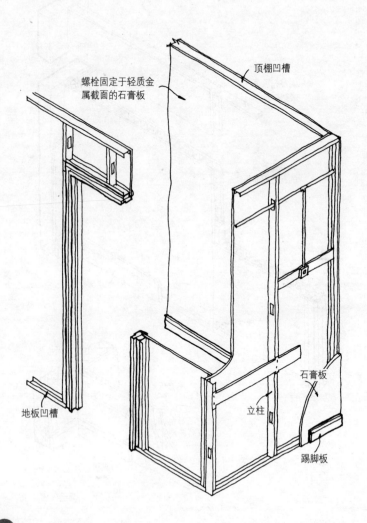

- 螺栓固定于轻质金属截面的石膏板
- 顶棚凹槽
- 地板凹槽
- 立柱
- 石膏板
- 踢脚板

隔墙

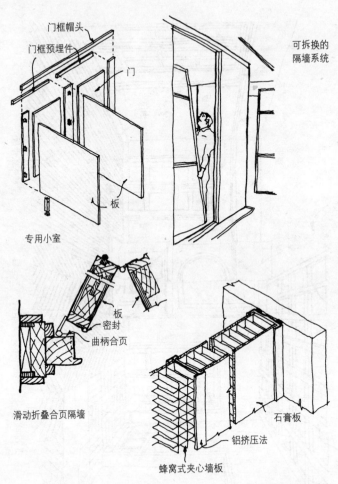

门框帽头
门框预埋件
门
板
专用小室

可拆换的隔墙系统

板
密封
曲柄合页
滑动折叠合页隔墙

石膏板
铝挤压法
蜂窝式夹心墙板
框架墙板隔墙系统

室内

4. 上部结构——屋盖

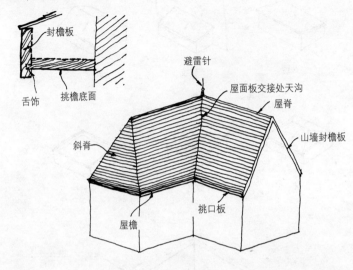

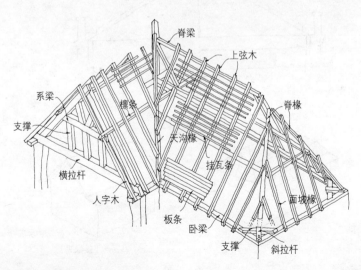

斜屋顶类型

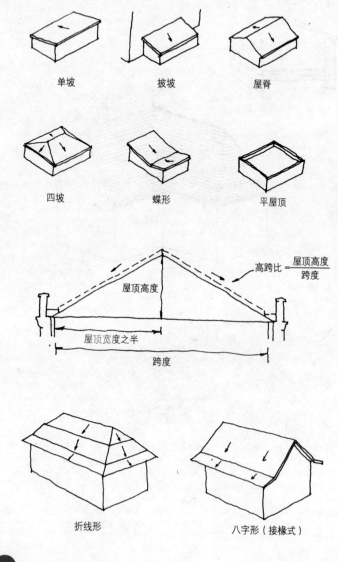

传统木屋顶类型

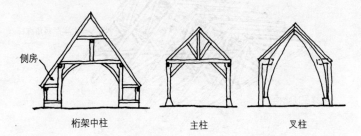

桁架中柱　　　　主柱　　　　叉柱

侧房

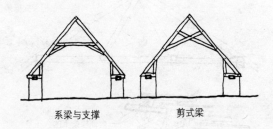

系梁与支撑　　　　剪式梁

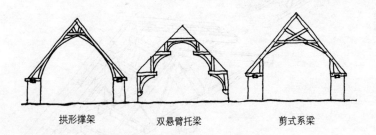

拱形撑架　　　　双悬臂托梁　　　　剪式系梁

桁架

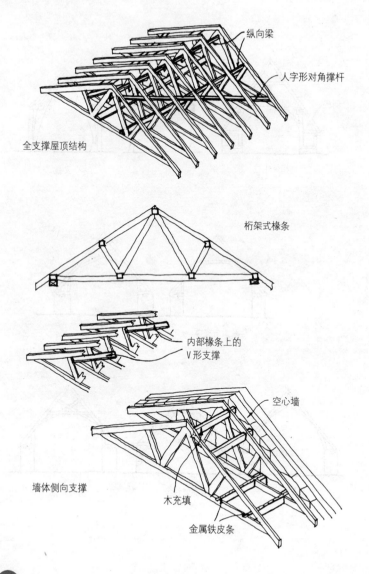

纵向梁

人字形对角撑杆

全支撑屋顶结构

桁架式椽条

内部椽条上的V形支撑

空心墙

墙体侧向支撑

木充填

金属铁皮条

桁架

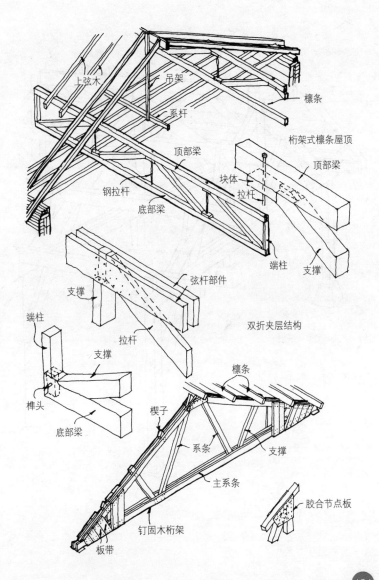

螺栓和桁架连接件

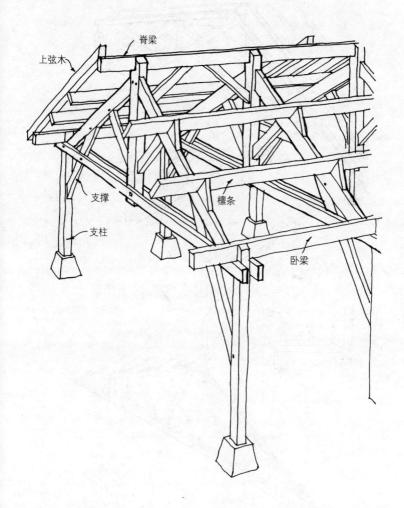

斜半屋架

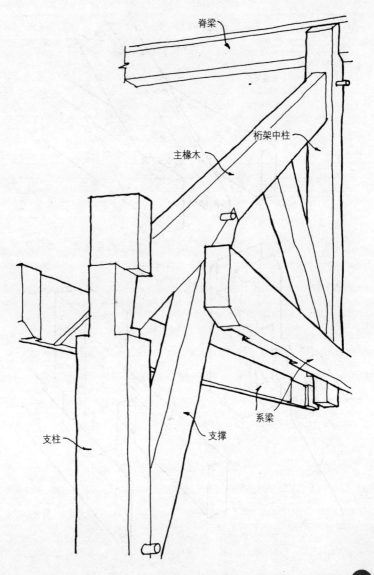

桁架橼条组装

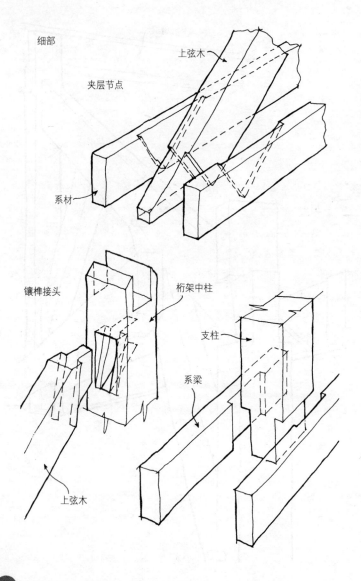

细部

上弦木

夹层节点

系材

镶榫接头

桁架中柱

支柱

系梁

上弦木

桁架椽条组装

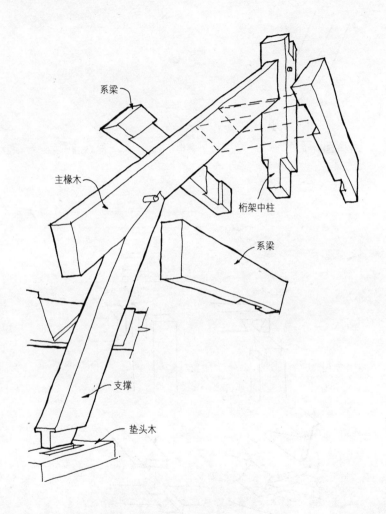

木结构节点

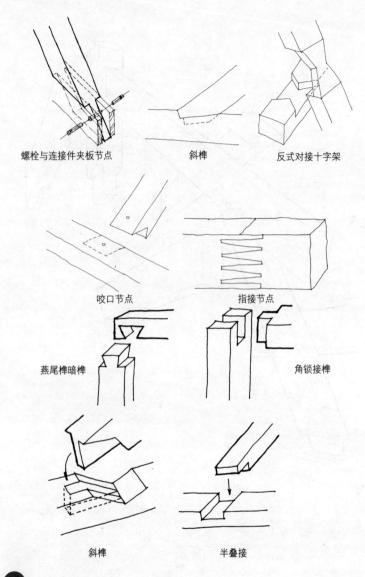

螺栓与连接件夹板节点　　斜榫　　反式对接十字架

咬口节点　　指接节点

燕尾榫暗榫　　角锁接榫

斜榫　　半叠接

交叉木节点

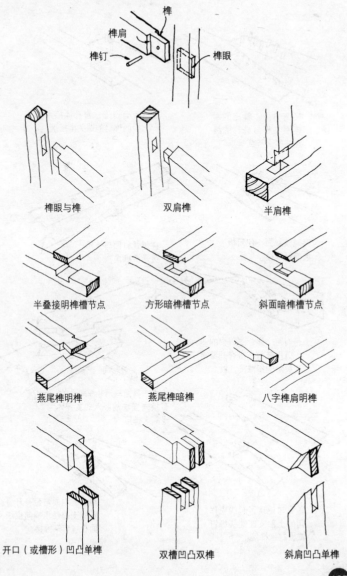

传统嵌接节点

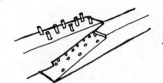

带有方形、斜孔的端接面与板面榫钉的终止式斜面节点

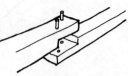

带有方形、直孔端接面与板面榫钉的端头半搭接

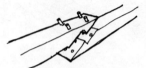

穿越式斜面与使用板面榫钉榫接

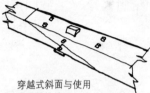

穿越式斜面与使用贯通榫榫接

带有两个终止式斜面的四部分斜嵌槽与带有开着斜孔突沿的扣榫接合板

终止式斜面与用开着斜孔的，有横向楔块的嵌入式支座木橡头来榫接

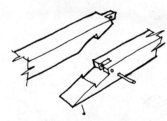

斜面与使用可起约束作用的上部支座端头榫钉或面钉来榫接

终止式斜面与带有开着斜孔的支座使用横向楔块及面板榫钉来榫接

传统嵌接节点

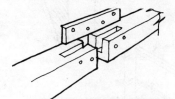

带有方形与竖直端接面的，三部分的鱼尾夹板式嵌接

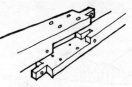

端头半搭接与带有托梁式支座的终止式斜面嵌接

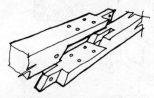

端头半搭接与带有开着斜孔的支座榫接

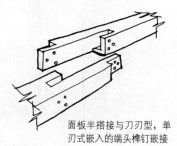

面板半搭接与刀刃型，单刃式嵌入的端头榫钉嵌接

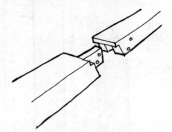

带有开着斜孔支座，重叠面板，端头榫钉，四分之三深的直式榫接

交叉刃式，用端头榫钉进行面板半搭接

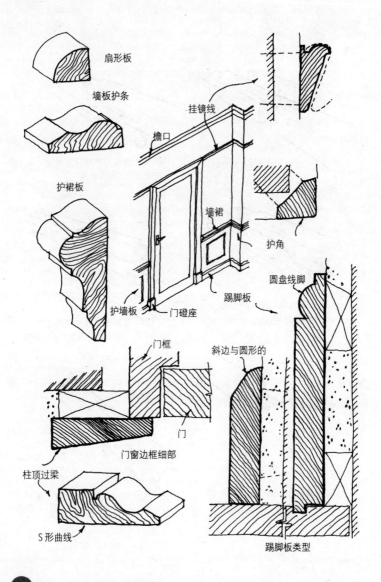

平屋顶

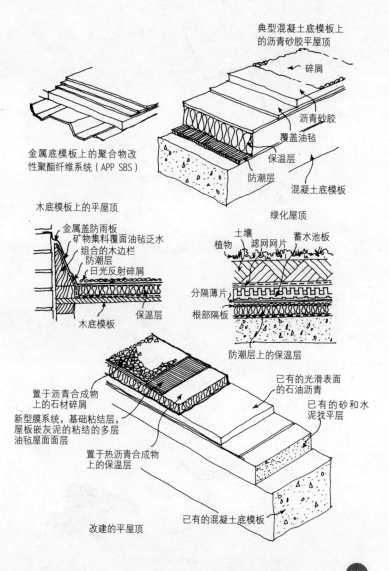

屋面覆盖层类型

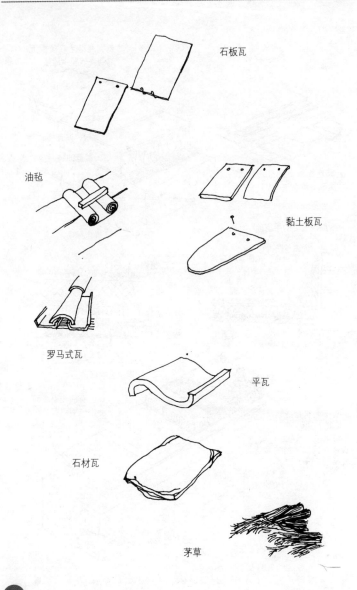

石板瓦

油毡

黏土板瓦

罗马式瓦

平瓦

石材瓦

茅草

石板瓦

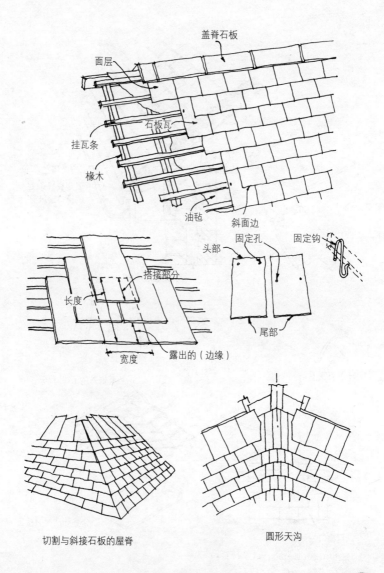

切割与斜接石板的屋脊

圆形天沟

平瓦

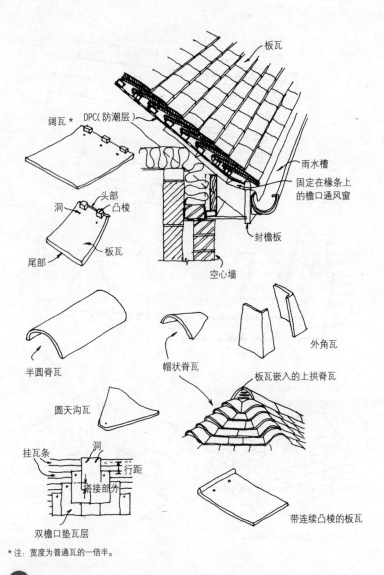

*注：宽度为普通瓦的一倍半。

波形瓦

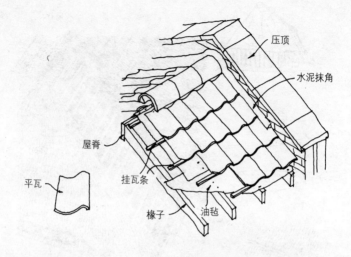

平瓦 / 屋脊 / 挂瓦条 / 椽子 / 油毡 / 压顶 / 水泥抹角

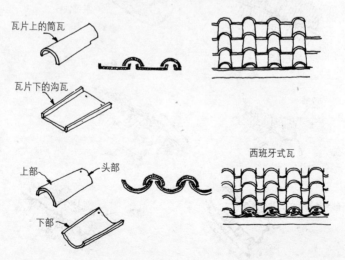

罗马（或意大利）式瓦

瓦片上的筒瓦
瓦片下的沟瓦

西班牙式瓦

上部 / 头部 / 下部

茅草屋顶

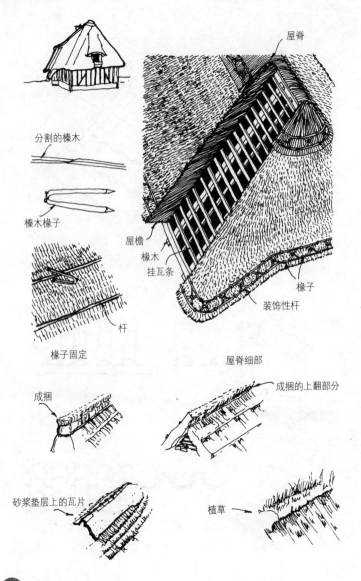

分割的榛木

榛木椽子

椽子固定

杆

屋脊
屋檐
椽木
挂瓦条
椽子
装饰性杆

屋脊细部

成捆

成捆的上翻部分

砂浆垫层上的瓦片

植草

传统木屋面板

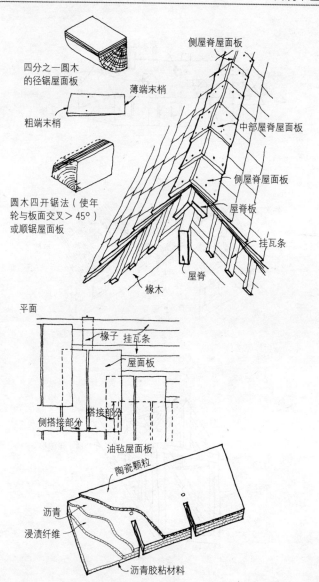

屋盖开洞

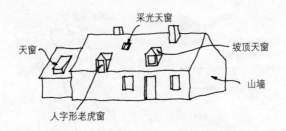

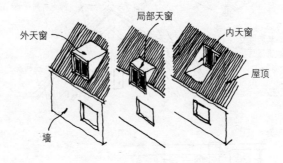

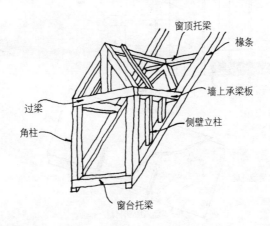

屋顶窗

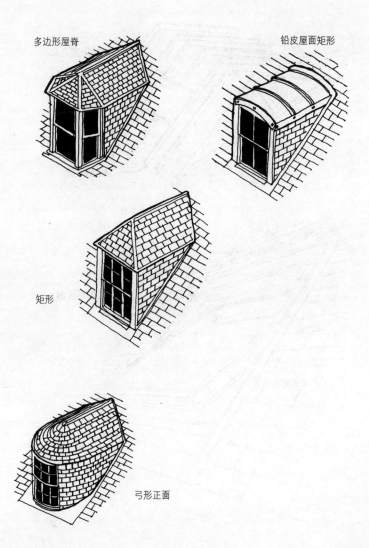

多边形屋脊

铅皮屋面矩形

矩形

弓形正面

天窗

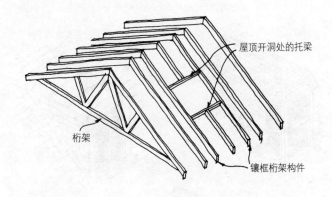

屋顶开洞处的托梁

桁架

镶框桁架构件

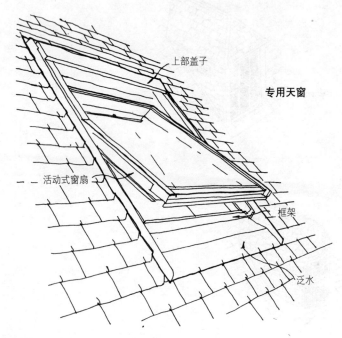

上部盖子

专用天窗

活动式窗扇

框架

泛水

烟囱

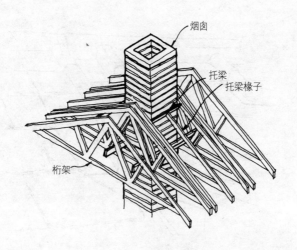

烟囱
托梁
托梁椽子
桁架

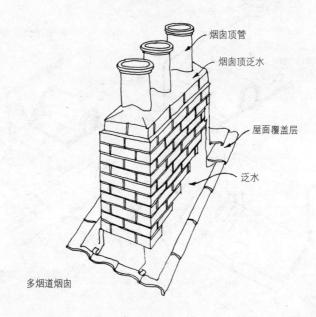

烟囱顶管
烟囱顶泛水
屋面覆盖层
泛水

多烟道烟囱

防风雨——铅皮

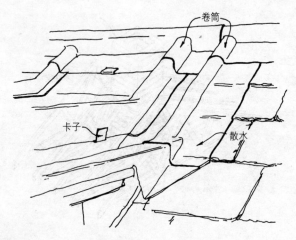

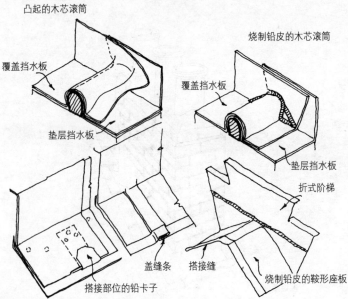

防风雨——挡雨板

屋顶/墙体连接点

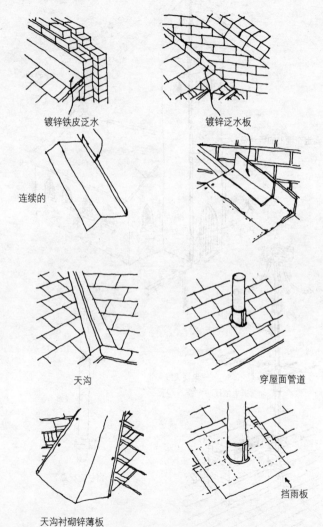

镀锌铁皮泛水

镀锌泛水板

连续的

天沟

穿屋面管道

天沟衬砌锌薄板

挡雨板

避雷装置

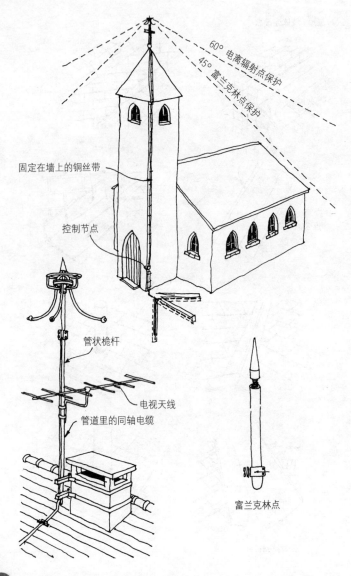

雨水管

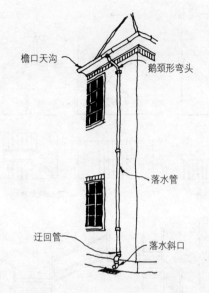

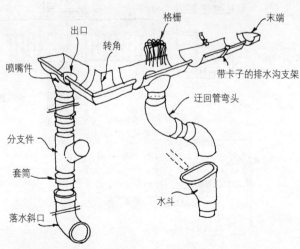

楼梯类型

5.上部结构——楼梯

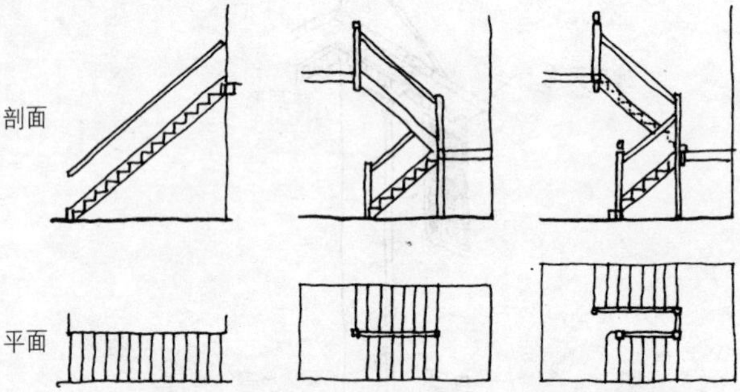

剖面

平面

直跑楼梯　　双折楼梯　　露明梯井楼梯（三跑楼梯）

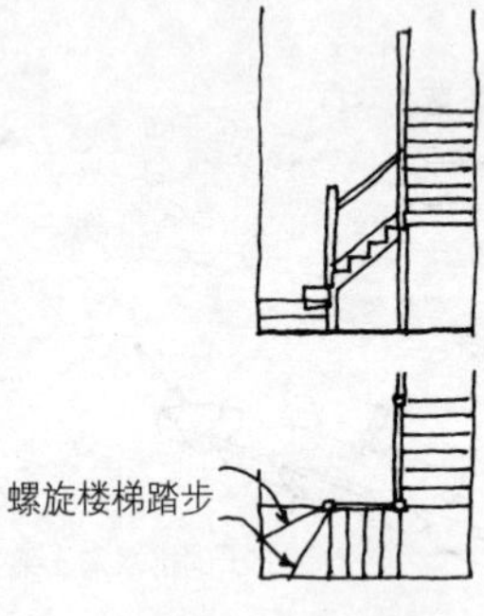

螺旋楼梯踏步

直角转弯楼梯 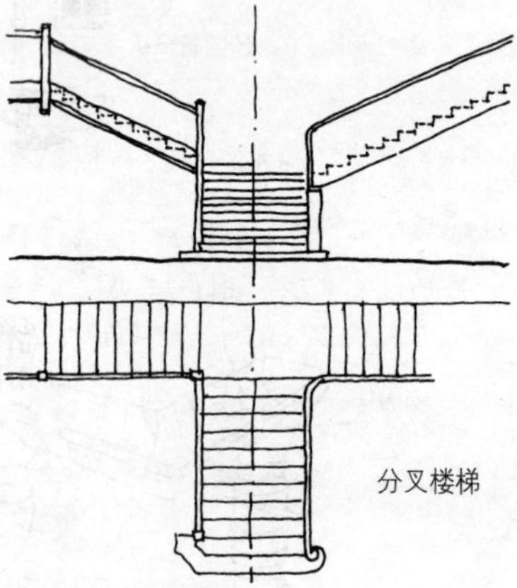　　　　　　　　　　分叉楼梯

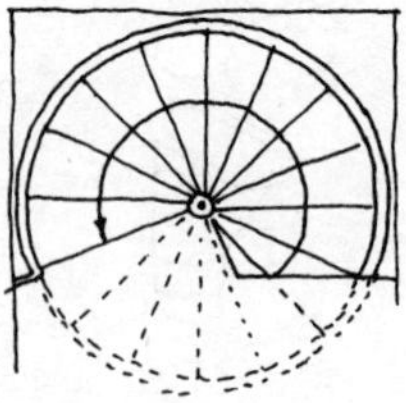

螺旋楼梯

传统木楼梯

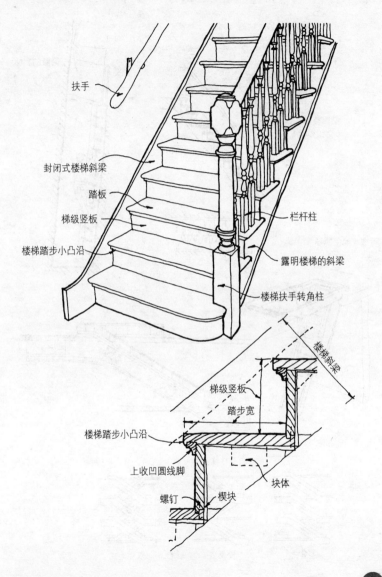

特殊楼梯和自动扶梯

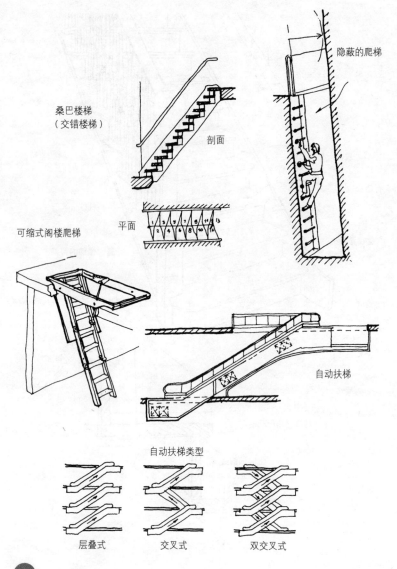

电梯

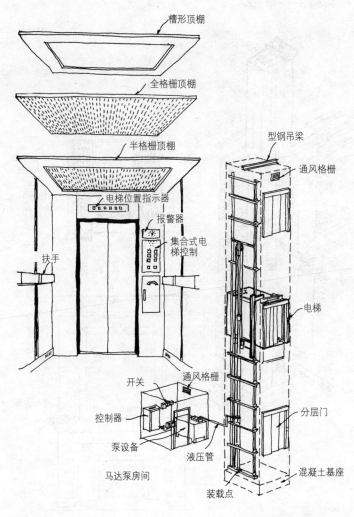

液压电梯井

壁炉

6.上部结构——烟囱

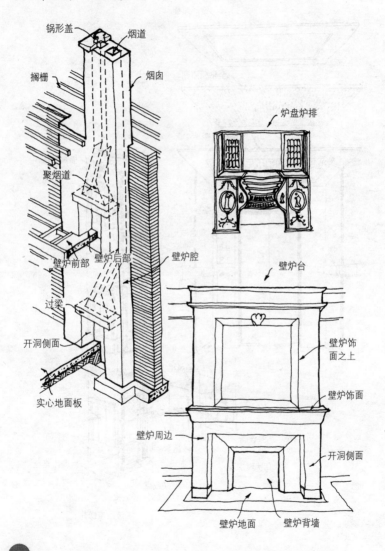

壁炉附件

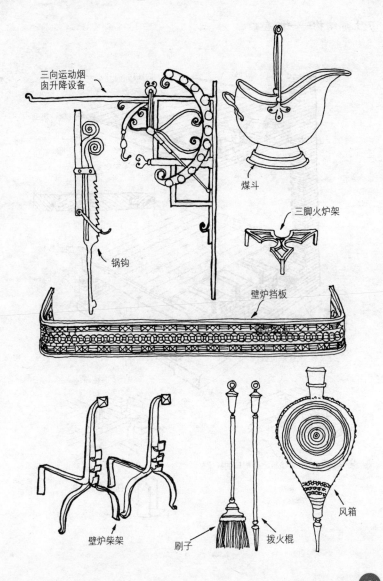

7.上部结构——楼盖

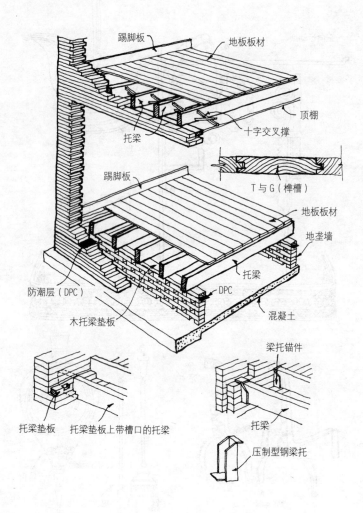

木楼盖——饰边开洞

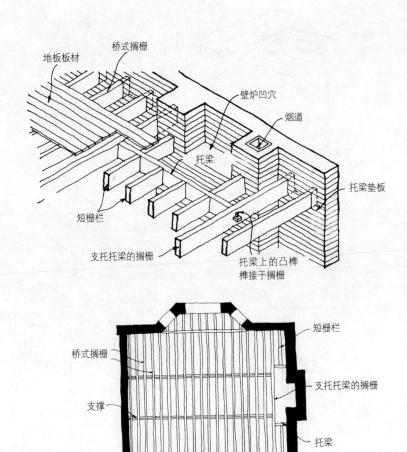

单层托梁地板平面

木楼盖镶边

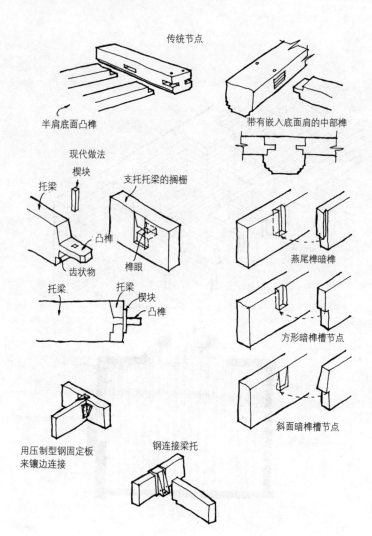

拼板与角接

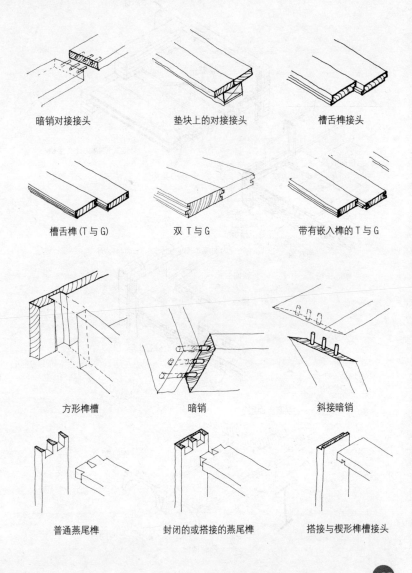

暗销对接接头　　　　　　垫块上的对接接头　　　　　槽舌榫接头

槽舌榫(T与G)　　　　　　双T与G　　　　　　　　带有嵌入榫的T与G

方形榫槽　　　　　　　　暗销　　　　　　　　　　斜接暗销

普通燕尾榫　　　　　　　封闭的或搭接的燕尾榫　　　搭接与楔形榫槽接头

混凝土楼盖

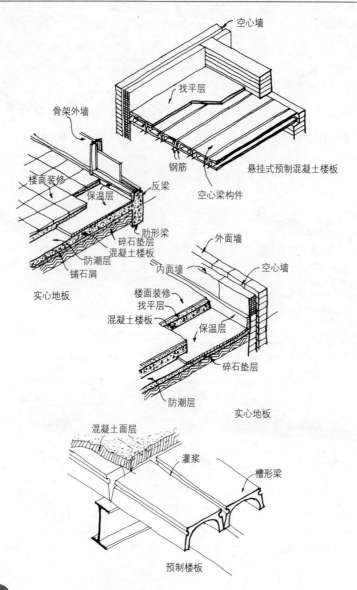

钢筋混凝土楼盖

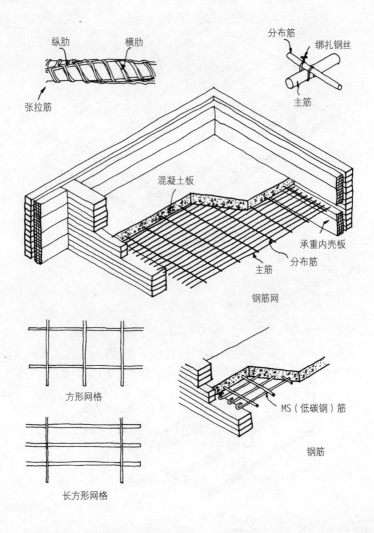

顶棚——木托梁

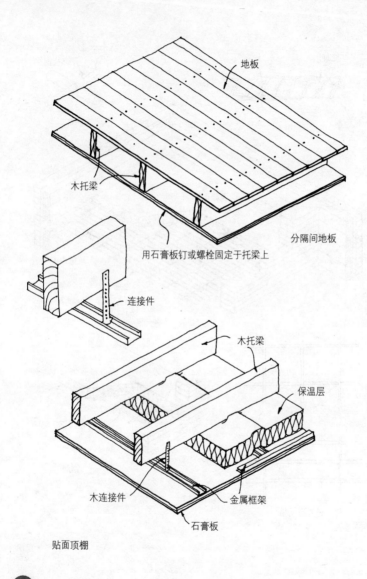

- 地板
- 木托梁
- 分隔间地板
- 用石膏板钉或螺栓固定于托梁上
- 连接件
- 木托梁
- 保温层
- 木连接件
- 金属框架
- 石膏板

贴面顶棚

吊顶

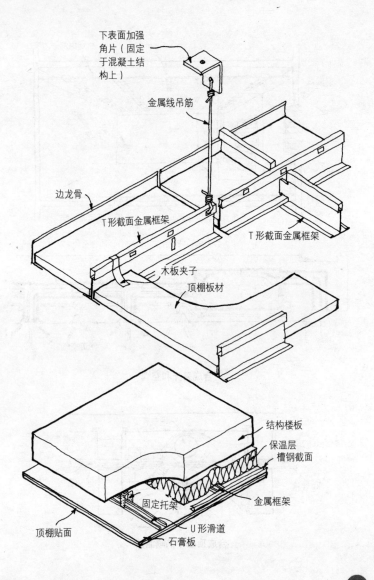

设备空间

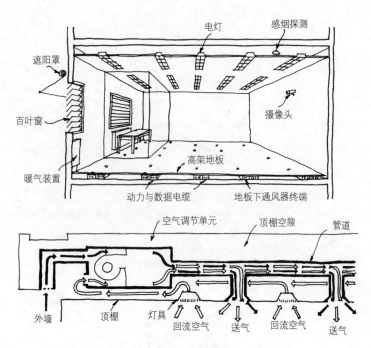

合成采光和空调

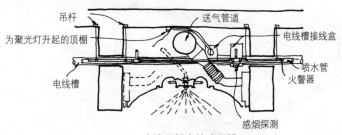

合成顶棚中的喷洒器

高架地板

浅空隙板条楼板

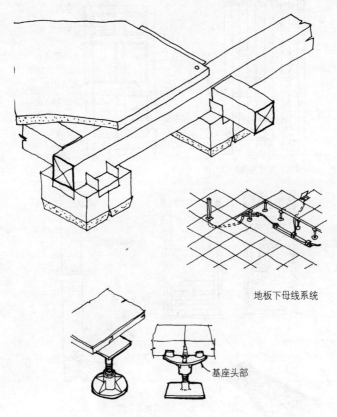

地板下母线系统

基座头部

深空隙平台楼板支架

8.上部结构——墙上开洞

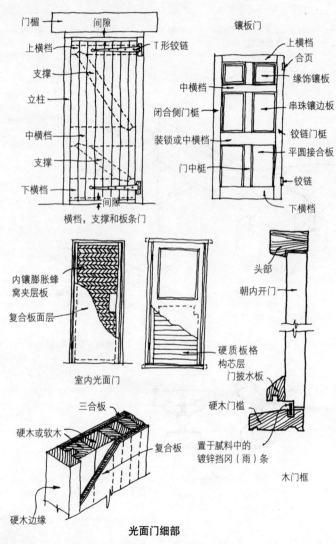

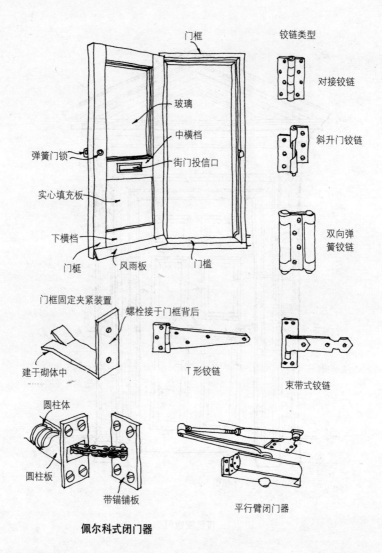

门

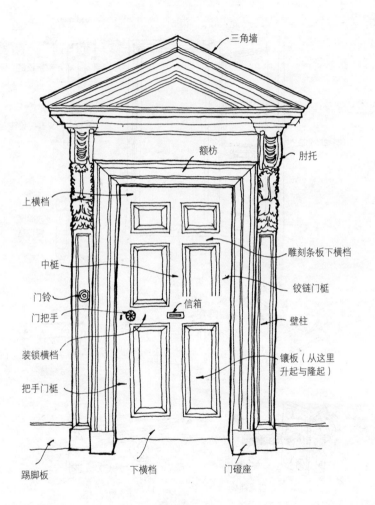

传统镶板门

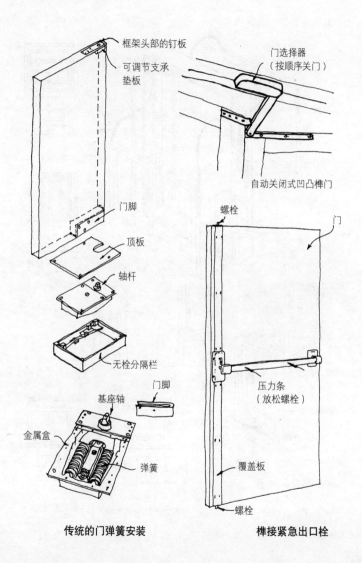

传统的门弹簧安装 **榫接紧急出口栓**

传统窗户——格窗

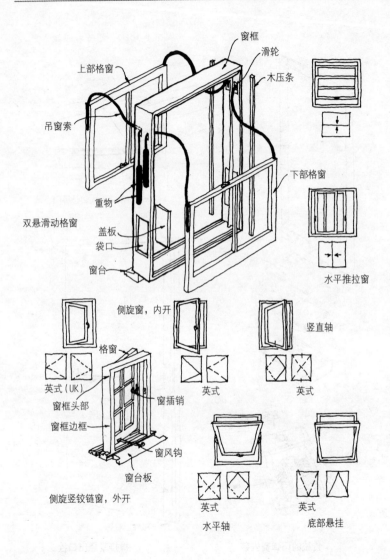

传统窗户——竖铰链窗

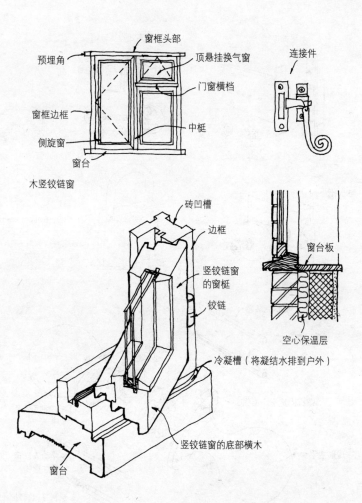

木竖铰链窗

建筑五金——把手与门锁

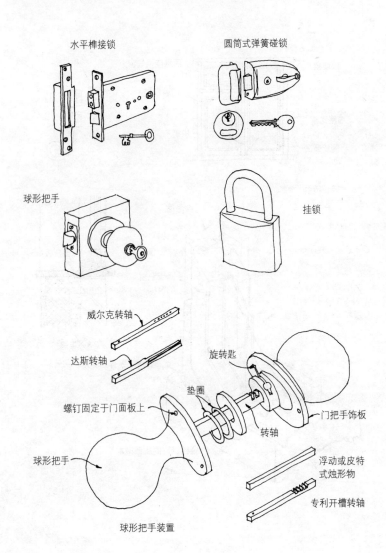

建筑五金——螺栓和把柄

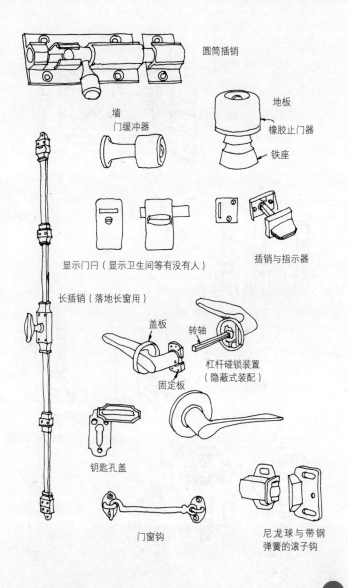

建筑五金——锁

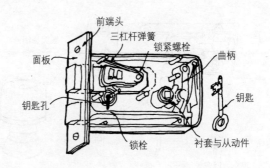

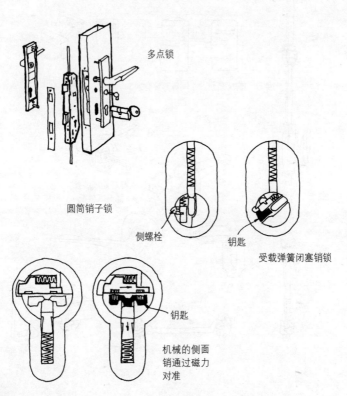

建筑五金——铰链

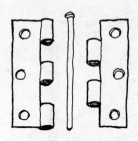
抽芯铰链

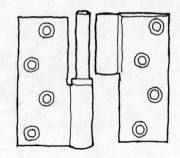
活脱铰链

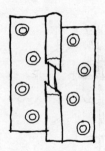
斜升门铰链（开门时，一页升高）

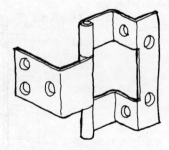
弯曲铰链

H形铰链

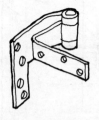
长翼铰链

151

9. 上部结构——附件

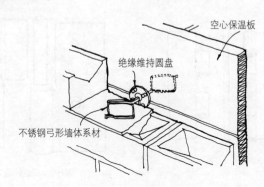

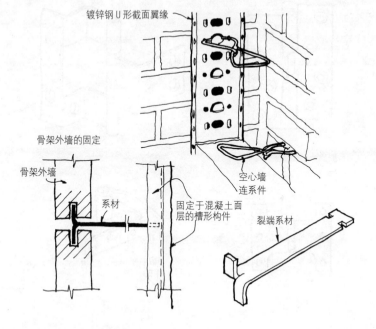

金属附件——搁栅锚件

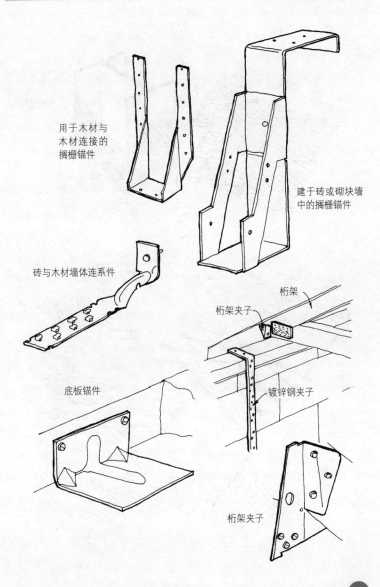

用于木材与木材连接的搁栅锚件

建于砖或砌块墙中的搁栅锚件

砖与木材墙体连系件

桁架

桁架夹子

底板锚件

镀锌钢夹子

桁架夹子

金属部件——拉展金属网

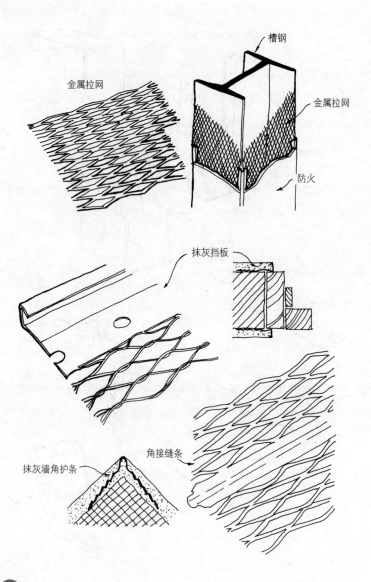

金属部件——过梁

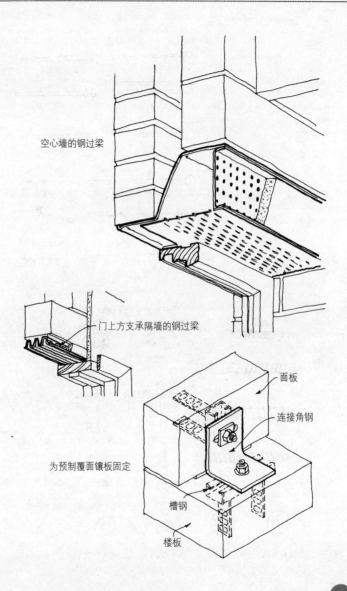

空心墙的钢过梁

门上方支承隔墙的钢过梁

为预制覆面镶板固定

面板

连接角钢

槽钢

楼板

钉子与螺钉

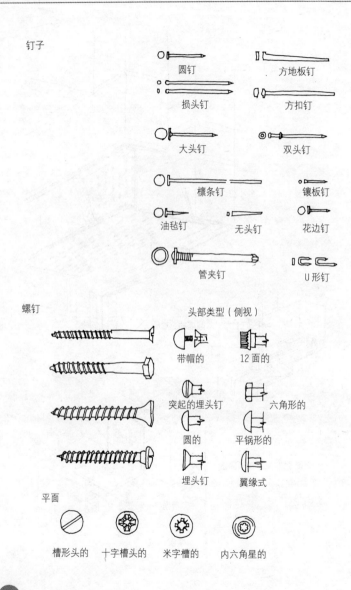

螺栓和栓头

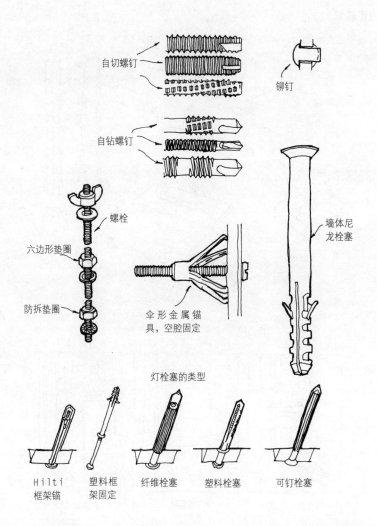

灯栓塞的类型

10.装修

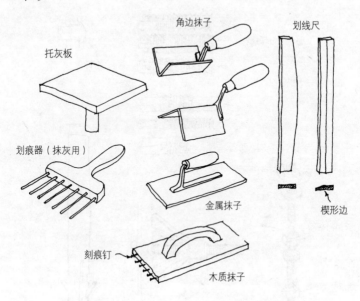

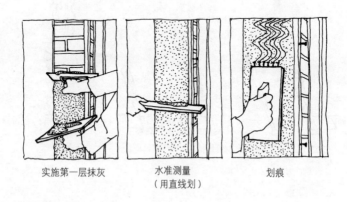

实施第一层抹灰 　　水准测量（用直线划）　　划痕

抹灰篱笆墙

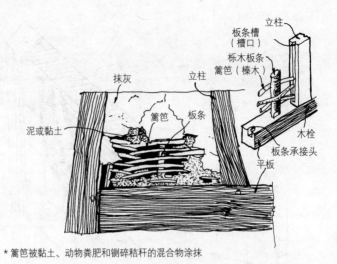

*篱笆被黏土、动物粪肥和铡碎秸秆的混合物涂抹

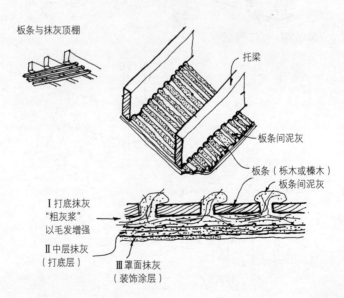

粉饰灰泥

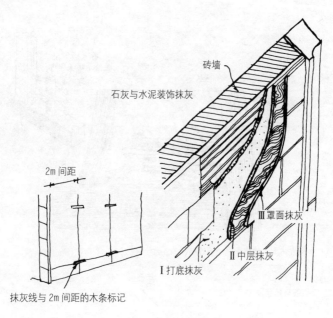

2m 间距

抹灰线与 2m 间距的木条标记

砖墙

石灰与水泥装饰抹灰

Ⅲ 罩面抹灰
Ⅱ 中层抹灰
Ⅰ 打底抹灰

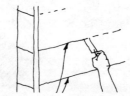

在罩面抹灰还软时刻出的模缝线

垂直抹灰分隔条

打底层

装饰性抹灰泥作业

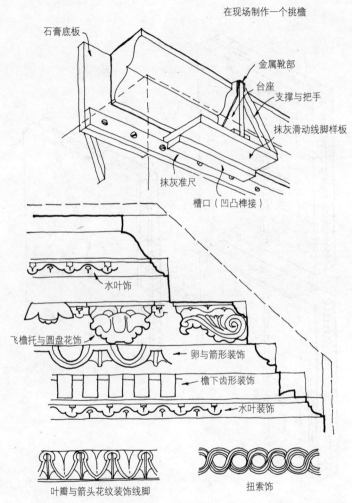

檐口装饰（置于湿的装饰抹灰中）

内墙——干衬砌

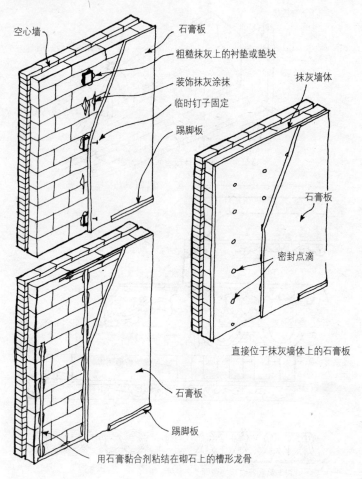

直接固定于墙上的石膏板

- 空心墙
- 石膏板
- 粗糙抹灰上的衬垫或垫块
- 装饰抹灰涂抹
- 临时钉子固定
- 踢脚板

直接位于抹灰墙体上的石膏板

- 抹灰墙体
- 石膏板
- 密封点滴

固定在轻质金属槽形副龙骨上的石膏板

- 石膏板
- 踢脚板
- 用石膏黏合剂粘结在砌石上的槽形龙骨

木装修

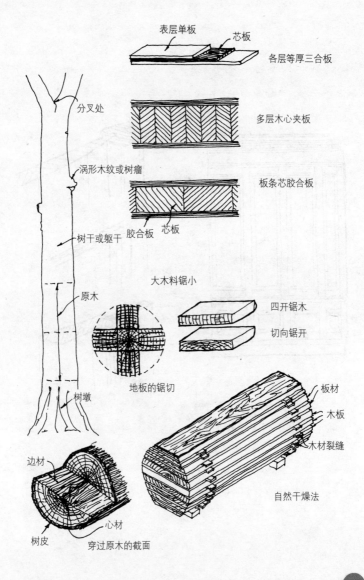

镶板

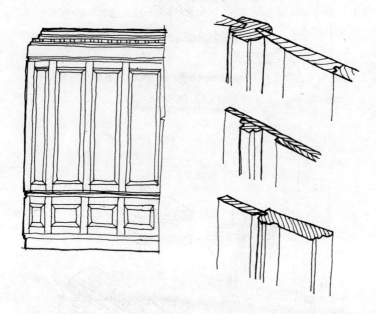

地毯

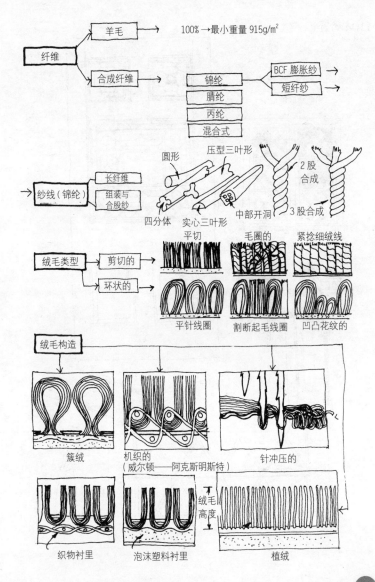

11.玻璃窗

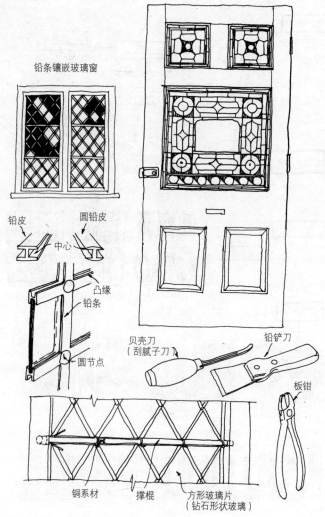

铅条镶嵌玻璃窗

铅皮　圆铅皮　中心　凸缘　铅条　圆节点

贝壳刀（刮腻子刀）　铅铲刀　板钳

铜系材　撑棍　方形玻璃片（钻石形状玻璃）

平板玻璃

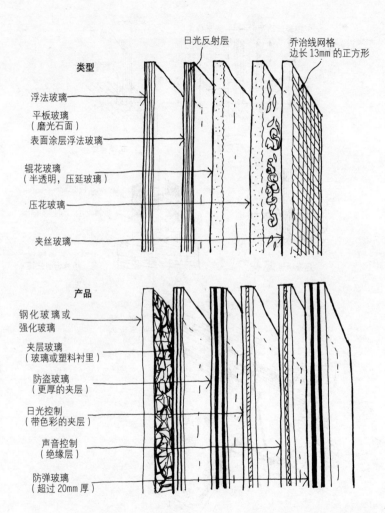

类型
- 浮法玻璃
- 平板玻璃（磨光石面）
- 表面涂层浮法玻璃
- 辊花玻璃（半透明，压延玻璃）
- 压花玻璃
- 夹丝玻璃

日光反射层

乔治线网格 边长13mm 的正方形

产品
- 钢化玻璃或强化玻璃
- 夹层玻璃（玻璃或塑料衬里）
- 防盗玻璃（更厚的夹层）
- 日光控制（带色彩的夹层）
- 声音控制（绝缘层）
- 防弹玻璃（超过20mm厚）

167

玻璃安装系统

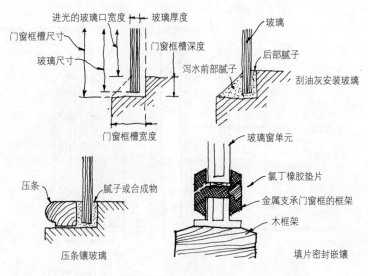

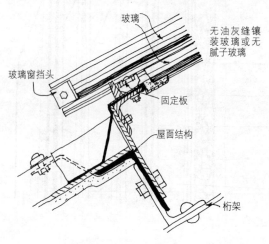

12.服务——排水与管道系统

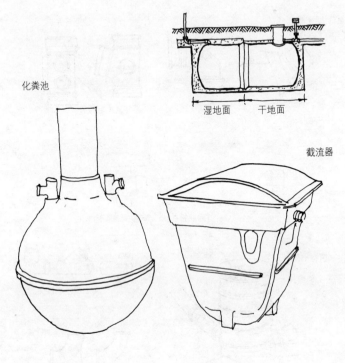

污水池

化粪池

湿地面　干地面

截流器

地面排水

通管孔（排水管分支端部的检查孔）

地下排水

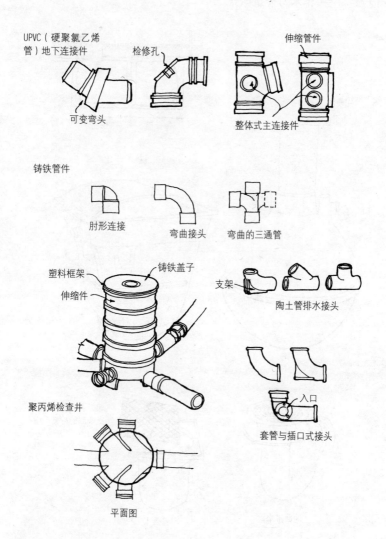

管道系统、给水排水系统

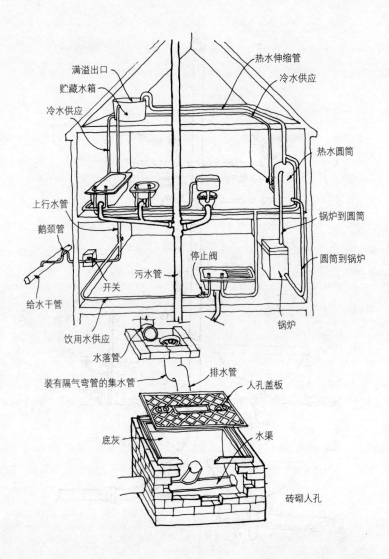

卫生管道作业

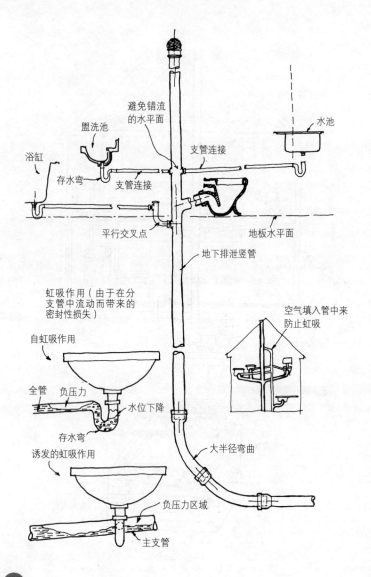

冲洗式坐式便桶与储水箱

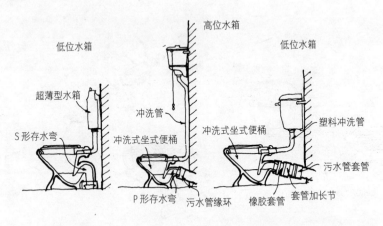

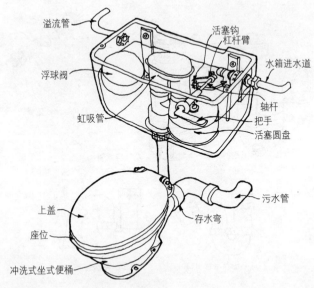

管道系统连接

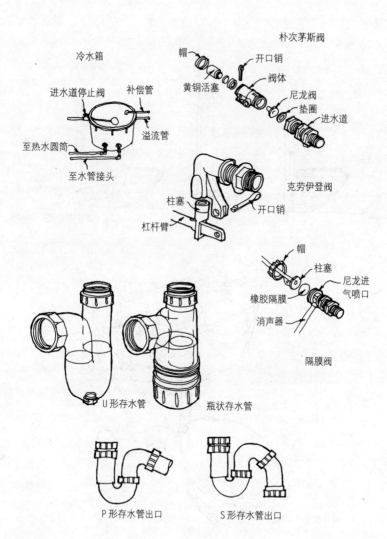

水加热

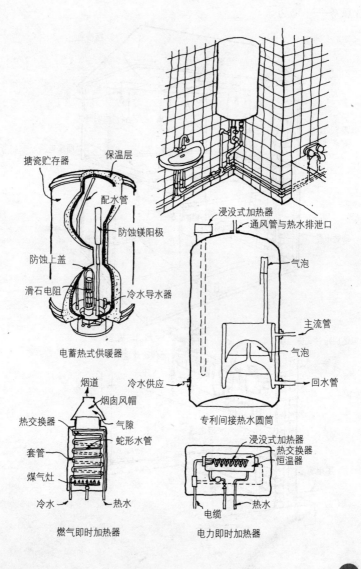

电蓄热式供暖器

燃气即时加热器

专利间接热水圆筒

电力即时加热器

13. 服务——电力

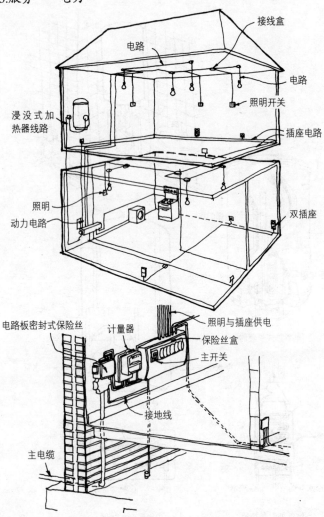

电力——附件

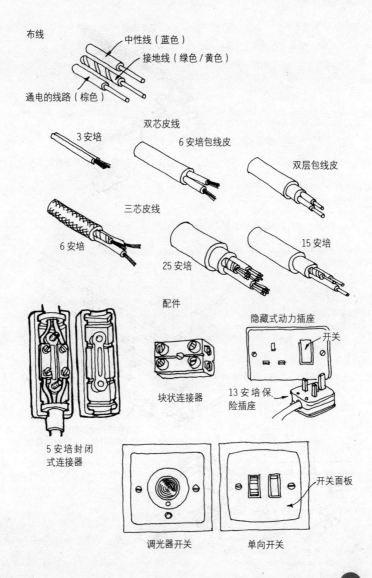

电力——布线路线

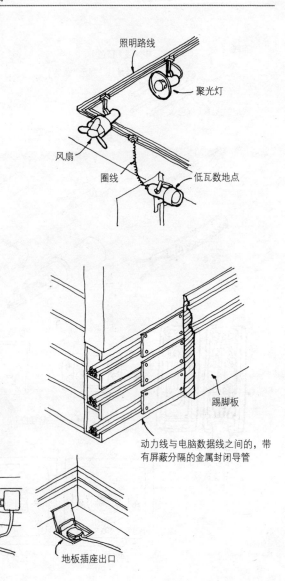

照明路线
聚光灯
风扇
圈线
低瓦数地点
踢脚板
动力线与电脑数据线之间的，带有屏蔽分隔的金属封闭导管
动力路线
安全插头
地板插座出口

户外照明

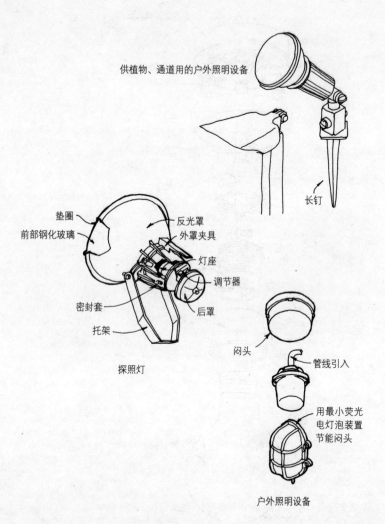

供植物、通道用的户外照明设备

长钉

垫圈　反光罩
前部钢化玻璃　外罩夹具
灯座
调节器
密封套
后罩
托架

探照灯

闷头
管线引入
用最小荧光电灯泡装置节能闷头

户外照明设备

电力装置

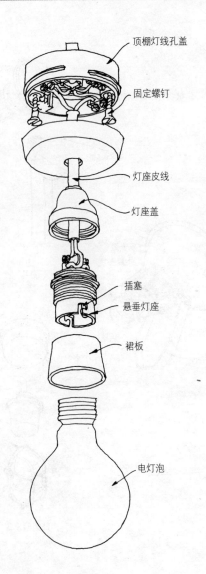

电力——灯泡

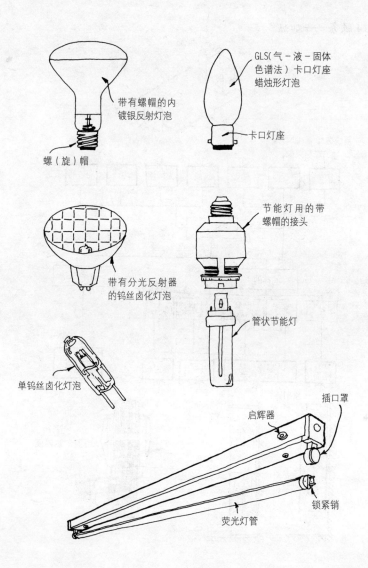

加热系统

14.服务——加热

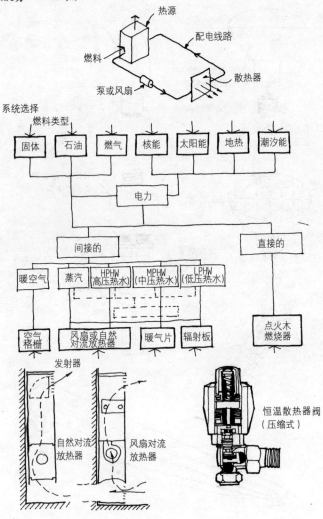

中央暖气系统——热水

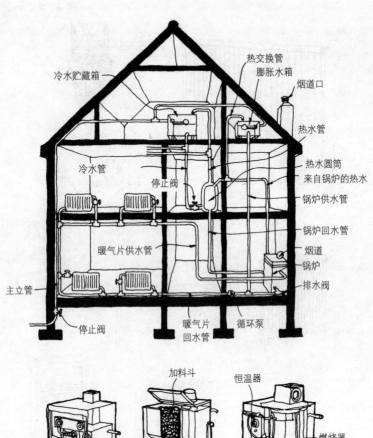

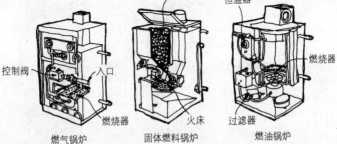

散热器

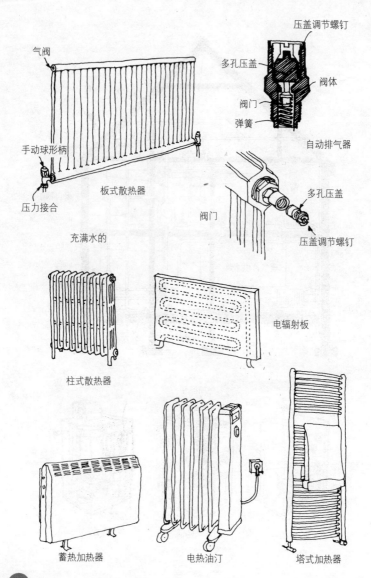

空调

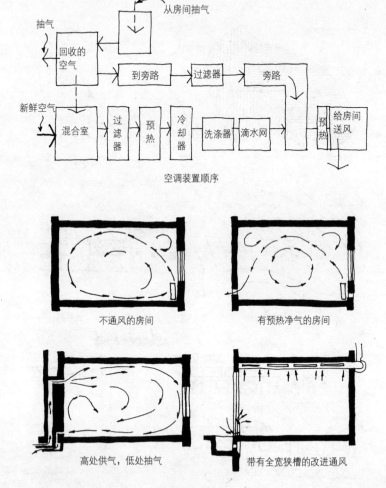

空调装置顺序

不通风的房间

有预热净气的房间

高处供气，低处抽气

带有全宽狭槽的改进通风

15. 户外作业/景观

土方工程的图形表述

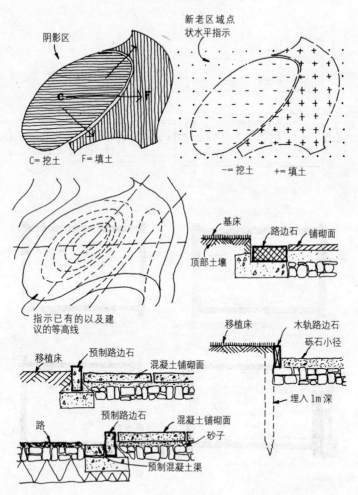

灌溉与屏蔽

灌溉系统

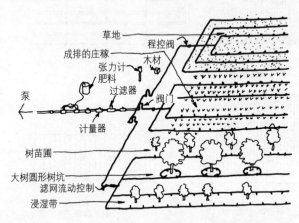

屏蔽

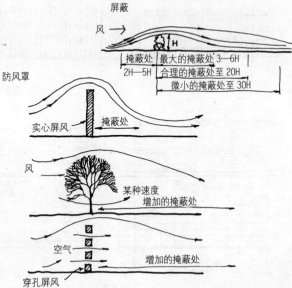

围栏

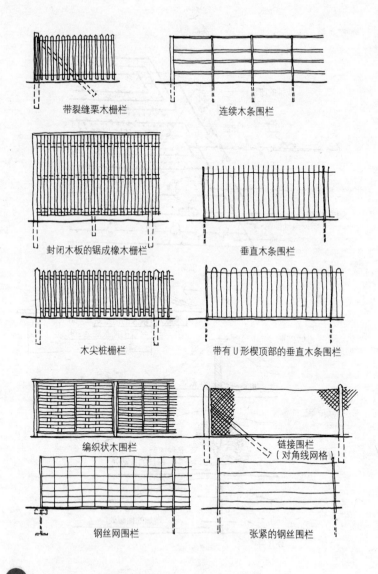

树木

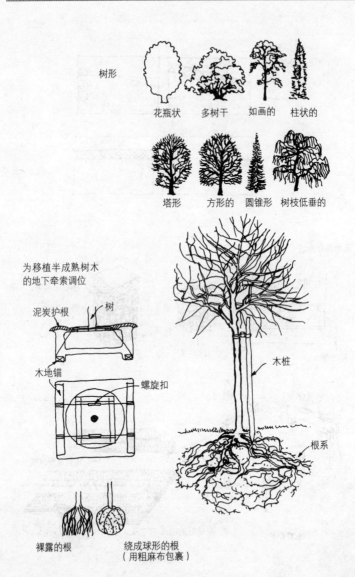

土地平整、草皮铺设

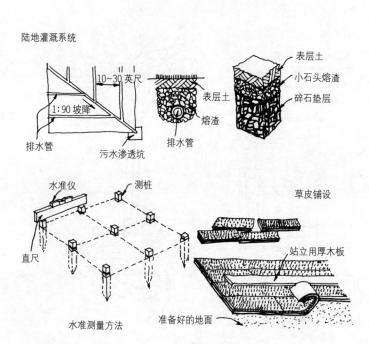

植物养护、附属房屋

格架

园艺用钟形玻璃盖

苗床罩子

温室台架

温室

凉亭

温室

第6章 环境状况

全球变暖与温室效应　195

可持续绿色建筑　196

危险中的建筑:自然灾害　201

全球变暖与温室效应

1.全球变暖与温室效应

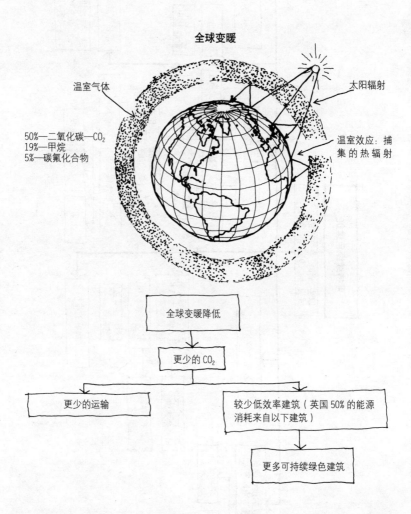

2.可持续绿色建筑

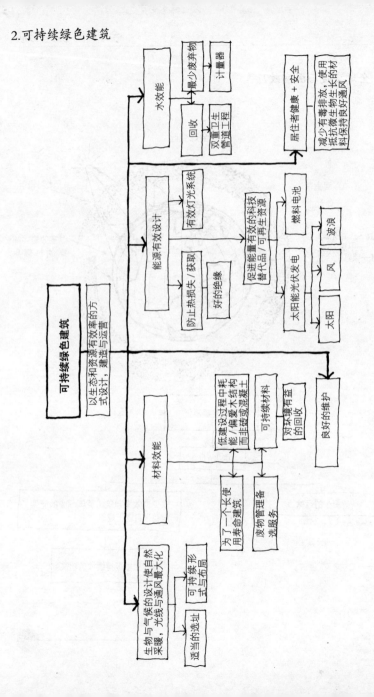

有关生物与气候的设计

传统建筑

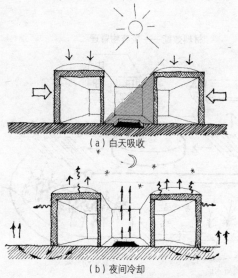

(a) 白天吸收

(b) 夜间冷却

现代建筑

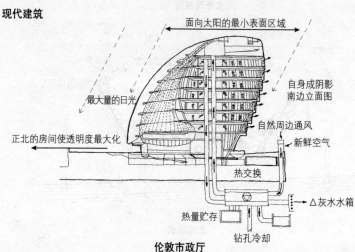

伦敦市政厅

注：△灰水，是指住宅室内排水中未被粪便污染的部分，包括洗衣、沐浴、盥洗以及厨房等污水。

材料功效——废物管理

材料效能——废弃物管理

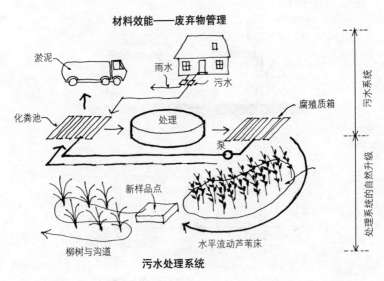

污水处理系统

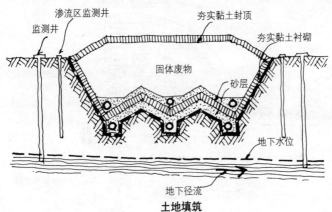

土地填筑

能量效能设计

能量效能设计：能量的替代来源

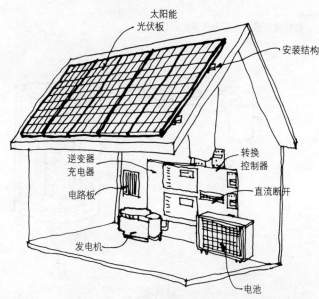

混合物动力系统

能量效能设计：能量来源的替代物

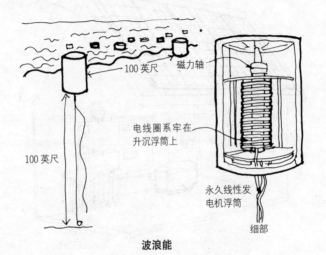

100 英尺
磁力轴
电线圈系牢在升沉浮筒上
100 英尺
永久线性发电机浮筒
细部

波浪能

空气排出
发电机
涡轮机
波浪方向
操作原理

3.危险中的建筑：自然灾害

特性

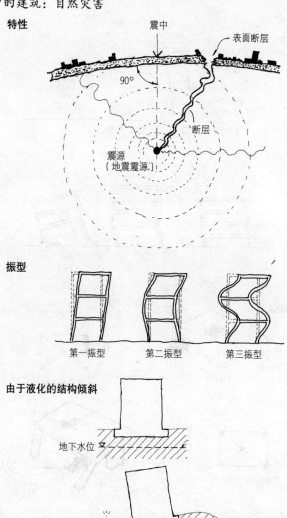

振型

第一振型　　第二振型　　第三振型

由于液化的结构倾斜

地下水位

水压力

地震

平坠

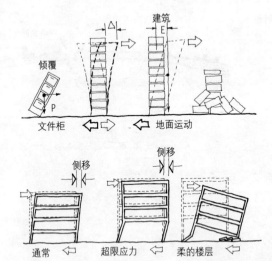

倾覆　文件柜　建筑　地面运动　侧移　通常　超限应力　柔的楼层

扭转效应

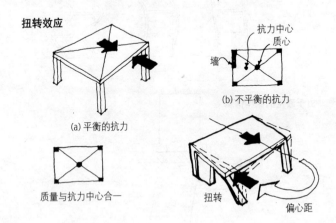

(a) 平衡的抗力

抗力中心　质心　墙

(b) 不平衡的抗力

质量与抗力中心合一　　扭转　偏心距

地震

剪切破坏

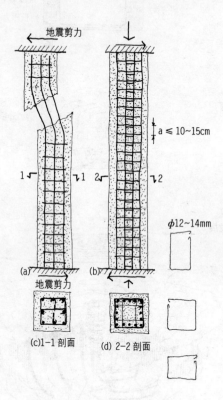

极端天气：原理，飓风

热水循环原理

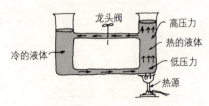

海风原理

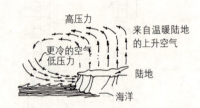

飓风的发展

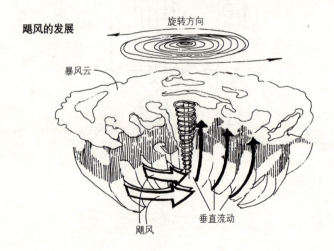

极端天气：飓风，闪电

极端天气：飓风，风效应

建筑上的风效应

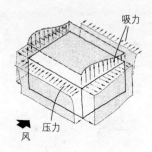

屋顶形状的风效应

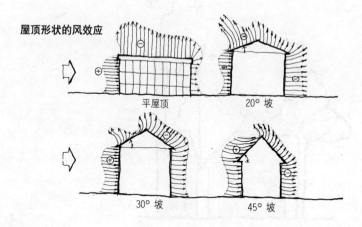

平屋顶　　　20°坡

30°坡　　　45°坡

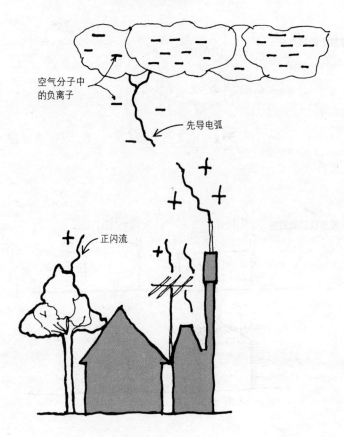

块体运动：生成，滑坡

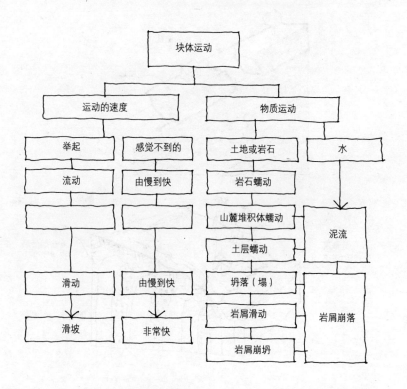

块体运动：生成，滑坡

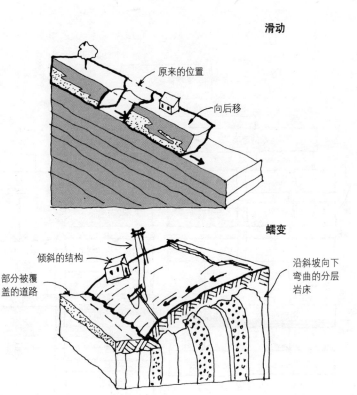

块体运动：生成，滑坡

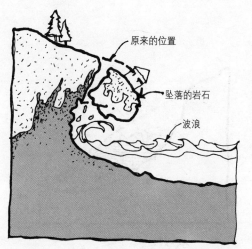

坠落

原来的位置
坠落的岩石
波浪

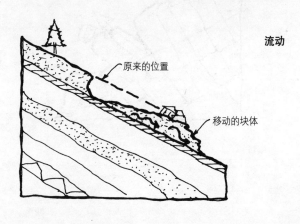

流动

原来的位置
移动的块体

块体运动：生成，滑坡

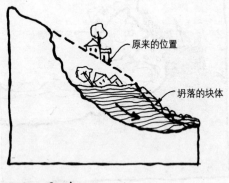

原来的位置
坍落的块体

坍落（塌）

倾覆

块体运动：生成，滑坡

隆起效应

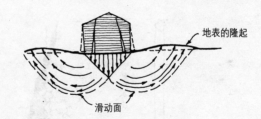

不均匀地面的不稳定性对建筑物的影响

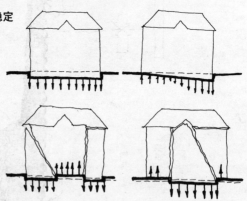

典型的由于地面影响造成的结构破坏

洪水

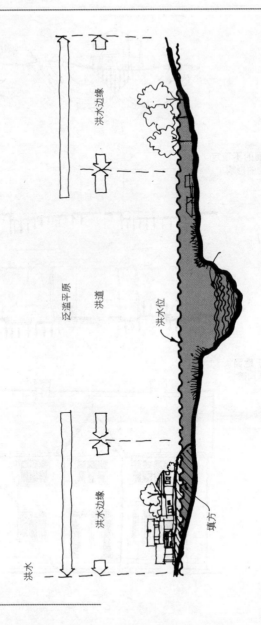

火山

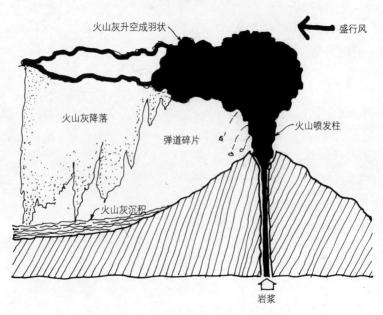

英汉专业词汇对照

Abutment 拱座，83
Account(final) 竣工结算，51
Accuracy 准确性，58
Accurate survey 准确的测量，58
Acroterion 山墙饰，16
Acts 法案，46
Adaptor 接头，181
Air brush 喷枪，10
Air conditioning plant 空调装置，185
Air handling unit 空气调节单元，140
Aisle(church) 教堂侧廊，23，24
Aisle(cattle) 家畜走廊，31
Alter 圣坛，24
Alternative sources of energy 能量的替代来源，199
Ambulatory 回廊，35
Angle 转角，125
Angle bead 护角，110
Angle cleat 连接角钢，155
Angle joint 角接，135
Angle of obstruction 障碍物角度，42
Angle of acceptance 可接受角，42
Angle poise lamp 角度平衡灯，9
Appraisal (development) 评估（开发），49
Appraisal (financial) 评估（财务），32
Approved documents(building regulations) 被批准的文件（建筑规程），45
Approximate(quantities estimating) 近似（工程量估算），51
Apron flashing 挡雨板，123
Apse 半圆壁龛，24
Arcade(brick) 砖连拱廊，83
Arch(flat) 拱（平的），89
Arch(semicircular) 拱（半圆），89
Arch(Gothic) 拱（哥特式），24
Arch braced roof 拱形撑架屋顶，99
Architect 建筑师，52，53
Architrave 柱顶过梁，11，104
Arris 边棱，84
Ashlar 琢石，87，89
"A" sizes 图幅尺寸，7
Assembly truss 装配桁架，104
Assembly rafter 装配椽条，104
Astragal 圈线，18
Automatic pencil 自动铅笔，10

Axonometric projection 不等角投影，5

Baffle 挡板，92
Baluster 栏杆柱，127
Balustrade 栏杆，25
Ballast 压载，62
Balleting 颠簸运动，88
Ball valve 球形阀，173
Bar(reinforcement) 钢筋，137
Bar(pressure) 压力条，145
Barbican 外堡，20
Bark 树皮，163
Barn shapes 仓房形状，31，32
Barn(cruck framed) 仓房（曲木框架），22
Barn(threshing) 脱粒仓房，32
Barrel bolt 圆筒插销，149
Bartizan 箭塔，20
Base 基脚，83
Base bed 基床，86
Base line 基线，57
Basement(dry area) 地下室墙（干燥区域），77
Basic module 基本模型，7
Bat 半砖，84
Battens 挂瓦条，97
Batter 斜坡，88
Bay(cart) 运货马车隔间，32
Bayonet cap 卡口灯头，181
Bay open front 前开式开间，32
Bay window 凸窗，37
Beam compass 横杆圆规，10
Beam(threshing) 梁（脱粒），32
Beam(interior) 梁（内部的），96
Beam(vault) 梁（拱顶），24
Bellows 风箱，131
Bench mark 水准基点，4
Beveled(bat closer) 斜面（半砖），84
Beveled housed joint 斜面暗榫槽节点，107，134
Billet 错齿式粉刷线脚，19
Bills of quantity 工程量清单，51
Binder 系杆，101
Binding 结合，5
Bioclimatic design 有关生物与气候的设计，197
Birds mouth joint 咬口节点，106
Bitumen 沥青，117

Blade(scalpel) 刀片（手术刀），10
Blinding 垫层，74，75
Blockboard 板条芯胶合板，163
Blocking course 女儿墙，87
Block of flats 公寓楼，30
Boards(blockboard laminboard) 板条芯胶合板、多层木心夹板，163
Boarding 隔板，26
Boaste finish 修凿装饰，86
Boiler 锅炉，171，183
Bole 树干，163
Bolster 阔凿，70
Bolt(panic) 紧急出口栓，145
Bolts 螺栓，157
Bolt and connector truss 螺栓和桁架连接件，102，106
Bond (brick), 82
  (English, Flemish, rat trap, reverse, half bond, quarter bond, third bond, broken bond) 砌合（英国式、荷兰式、空斗墙、反向、半砌合、1/4砌合、1/3砌合、不规则砌缝），82
Boom 吊杆，63，101
Bore hole cooling 钻孔冷却，197
Bossed lead roll 突起的木芯滚筒，122
Boundary line 边线，41
Bow window 弓形窗，37
Box spanner 套筒扳手，69
Brace 曲柄、支撑，69，142
Braced roof 支撑屋顶，100
Bracing: 支撑：
  chevron V形，100
  diagonal 对角，100
  longitudinal 纵向，100
bracket(gutter) 排水沟支架，125
bradawl 打眼钻，70
branch piece 分支件，125
brattice 木结构防卫工事，20
Braye 城墙垛，20
Break line 打断线，8
Brick 砖：
  arcade 连拱，83
  bonding 砌合，82
  footing 墙脚，80
  joints 接缝，81
  types 类型，84
  wall 墙，80

Brickwork bonding 砖砌合，82
Brief 短暂，52
Broached finish 钻孔的装饰，86
Building contracts 建筑承包合同，54
Building fabric 建筑构成，71
Building line 建筑红线，41
Building regulations 建筑规程，45
Building site 工地，55
Building types: 建筑类型：
  detached 独立式，27，30
  high rise 高层，30
  large, high 大型、高层，46
  residential 居住建筑，27，28，29，30
  rural 乡村建筑，31
  semi-detached 半独立式，27
  terraced 连排式，27
  traditional farm 传统农场，32
Bulbs(light) 电灯泡，180，181
Bulldozer 推土机，63
Bull nose(double) 双圆角砖，84
Bundles(thatch) 成捆的茅草屋顶，116
Bungalow 平房，30
Burl 树瘤，163
Burlap(wrapping) 用粗麻布包裹，189
Burner 燃烧器，183
Burr 涡形木纹，163
Busbar system 母线系统，141
Bush and follower 衬套与从动件，150
Butt end 粗端末梢，117
Butterfly roof 蝶形屋顶，98
Butt joint 接头，135
Byre 牛棚，31，32

Cabin hook 门窗钩，149
Cable 缆索，19
Cable(mains electric) 主电缆，176
Cable looped 圈绕，178
Camera(metric) 公制照相机，58
Candle bulb 蜡烛形灯泡，181
Canopy(hipped, with cheeks) 顶棚（斜脊的、带有颊部的），31
Cap 柱帽，83
Capital 柱顶，16
Car 电梯，129
Carpentry joint 木结构节点，106
Carpet(pile,fibre,yarns) 地毯（绒毛、纤维、膨胀纱），165

Carriage 滑动托架, 9
Cart bay 运货马车隔间, 32
Casement window 竖铰链窗, 147
Cast iron fittings (knuckle joint, swept bent, tee) 铸铁管件(肘形连接、弯曲接头、弯曲的三通管) 170
Castle (mediaeval) 中世纪的城堡, 20, 29
Cavetto 凹圆线脚, 18
Cavity wall 空心墙, 26, 80, 114, 136, 162
Ceiling(level) 顶棚(水平面), 4
Ceiling(interior) 顶棚(室内), 96
Ceiling(lath and plaster) 顶棚(板条与抹灰), 159
Ceiling(timber) 顶棚(木的), 138
Ceiling(suspended) 吊顶, 139
Ceiling(liner) 顶棚(贴面), 138
Ceiling(rose) 顶棚(灯线孔盖), 180
Cellar 地窖, 25
Cellular core 蜂窝状夹心, 93
Central heating(hot water) 中央暖气系统——热水, 183
Centre line 点划线, 8
Certificate (making good defects, practical completion, final 执照(赔偿缺陷、实践完成、最终), 53
Cesspool 污水池, 169
Chamber 会议厅, 23
Chancel 高坛, 23
Change of use(material) 材料改变使用, 43
Channeled joint 沟槽形接缝, 81
Channels 沟槽形, 81, 162
Chantry 附属的小教堂, 23
Chapel 小教堂, 20, 23
Chateau 城堡, 29
Cheek 侧壁, 118
Chevet 多角室, 24
Chevron 人字纹, 19
Chimney breast 壁炉腔, 96, 130
Chimney crane 烟囱升降设备, 131
Chimney piece 壁炉台, 130
Chimney pot 烟囱顶管, 121
Chimney stack 烟囱, 25, 121, 130
Chimney 烟囱, 121
Chipping(solar reflective) 日光反射碎屑, 111
Chisel 凿子, 70
Choir 高坛, 24
Church(parish) 教区教堂, 23

Cill(stone) 石窗台, 87, 89
City of London Corporation 伦敦市的市政委员会, 43
Cistern(lower/high level, slim) 水箱(高低水位, 超薄型), 173
Cladding types 围护材料类型, 90, 91
Cladding fixing 骨架外墙的固定, 152
Clamp 夹钳, 69
Classical orders 古典柱式, 3, 17
Classical ornament 古典装饰, 18
Classical style 古典风格, 21
Classical temple 古典庙宇, 16
Claw hammer 羊角锤, 69
Claytiles 黏土瓦, 112
Cleat 楔子, 101
Clerestory 侧天窗, 23
Clerk of Works 工作人员, 53
Client 业主, 52, 53
Clip(lead) 铅卡子, 122
Clip(truss) 桁架夹子, 153
Cloche 园艺用钟形玻璃盖, 191
Closed string 封闭式楼梯斜梁, 127
Closer(brick) 接合砖, 84
Coal scuttle 煤斗, 131
Coat(backing,render,setting) 抹灰(打底层、打底、罩面), 158, 159
Coffer 顶棚嵌板, 96
Collar beam(root) 系梁(根部), 99
Collateral warranty 相关的保证, 54
Collective car control(lift) 混合式电梯控制, 129
Column 柱子, 16
Column base 柱基, 3, 16
Compartment floor 分隔间地板, 138
Component(working size) 组件的工作区尺寸, 4
Composite order 混合式柱式, 17
Compression gasket 压缩衬垫, 92
Compressor 压缩机, 67
Concrete floors(suspended, precast, solid) 混凝土楼板(悬挂式、预制、实心), 136
Concrete slab 混凝土板, 137
Concrete mixer 混凝土搅拌机, 66
Condensation 凝结, 85
Cone (truncated) 截头圆锥体, 6
Concerning(plumbing——valves, traps) 管道系统连接(阀、存水管), 174
Connector(fixing) 固定连接件, 138

Connector(electrical) 电路连接器,177
Conservatory 温室,185
Construction process 建设过程,47
Construction regulation(building acts) 施工规则（建筑法）,45
Consultants 顾问,52
Continuous foundation 连续基础,75
Contract 合同,51,53
Contractor 承包人,52
Controlling dimension 控制尺寸,4
Control: 控制:
  admin 管理,43
  construction 施工管理,45
  cost 成本,51
  Controls(continued) 控制
  legal 法律,46
  planning 规划,43
Convector(naturally, fan) 对流放热器（自然、风扇）,182
Convention(line) 常规线条,8
Conveyancing 产权转让,41
Coordination(dimensional) 尺度坐标,4
Coordinatograph 坐标仪,58
Coping 压顶,83,86,89,115
Corbel 托臂,20
Core(honeycomb) 蜂窝式夹心墙板,95
Corinthian order 科林斯柱式,17
Corner tower 角塔,20
Corn hole 谷物洞,32
Cornice 檐口,16,89,110
Cornice ornamentation 檐口装饰,161
Cornice(running in situ) 在现场制作一个挑檐,161
Corona 日冕线,18
Corner locking joint 角锁接榫,106
Cost(study, planning, check, monitoring, information) 成本（研究、规划、检查、监控、信息）,51
Cottage 小屋,27
Council(district, county, borough) 委员会（区、郡县、行政区）,43
Countersallied cross joint 反对接十字架节点,106
Courses(lacing) 拉结层,88
Covenant(restrictive) 限制性条款,41
Cover meter 面层测厚仪,60
Cover strip 盖缝条,92

Cramp(fixing) 固定夹紧装置,143
Crane 起重机,62
Cranked hinge 曲柄合页,95
Crenel 城垛口,20
Cross timber joints 交叉木节点,107
Crown(brick arch) 拱顶（拱砖）,83
Crown post roof 桁架中柱屋顶,99
Cruck:
  frame 曲木框架,22
  roof 叉柱屋顶,99
Cube 立方体,6
Cubicle(proprietary) 专用小室,95
Cupboard(built-in) 嵌入式壁橱,96
Curtain(castle) 城堡幕墙,20
Curtain wall(cladding) 幕墙覆层,91
Curve(French, flexible) 曲线尺、挠性曲线板,10
Cylinder(truncated) 截圆柱体,6
Cylinder locks(rim, latch) 圆筒销子锁、圆筒式弹簧碰锁,150,148
Cyma 波纹线脚,18
Cyma reversa 里反曲线,18

Dado 墩身、墙裙,16,96,110
Damp(effects) 潮湿的影响,85
Daylight 白昼,42
Daytime absorbtion 白天吸收,197
Deadmen(timber) 木桩柱,189
Debt financing 债务财务,50
Deck(timber for roof, metal, concrete) 底模板（用作屋顶的木的、金属的、混凝土的）,111
Deep void floor 深空隙楼板,141
Defects 缺陷,53
Defects liability period 缺陷责任期,53
Demountable partition 可拆换的隔墙,95
Dentils 檐下齿形装饰,161
Detached(building) 独立式建筑,27
Developer 开发商,50
Development:
  appraisal 开发评估,49
  funding 开发筹措资金,50
  permitted 许可的开发,43
Diameter(column base) 柱基的直径,3
Dichroic reflector 分光反射器,181
Dig and fill 开挖与回填,189
Dimension(controlling) 控制尺寸,4
Dimension(ratio of antiquity) 古代的尺寸比例,3

Dimensional coordination 尺寸坐标，4，7
Dimmer switch 调光器开关，177
Dipper arm 铲斗柄，63
District surveyor 地区工程检查员，46
Dog-leg(brick) 双折砖，84
Dog-leg(stair) 层间双折楼梯，126
Door 门：
　bell 门铃，144
　buffer 门缓冲器，149
　case 门框，33
　closers 闭门器，143，145
　hood 门罩，33
　knob 门把手，144
　lining 内门框，96
　selector 门选择，145
　spring 门弹簧器，145
　stop 止门器，149
Door types 门的种类，142，143，144，145
　Ledged and braced 横档与支撑，142
Door types(continued) 门的种类
　interior flush 室内光面，142
panelled 镶板，142，144
selfclosing 自动关闭式，145
Doorway 门道，33
Doric order 多立克柱式，17
Dormer 屋顶窗，25
Dormer windows (external, internal, gabled, partial, polygonal, piended, bowed, rectangular, lead roofed) 屋顶窗（外的、内的、人字形、局部、多边形、屋脊、弓形、矩形、铅皮屋面），112，119
Double bull nose 双圆角砖，84
Double cone 对顶锥，19
Double hammer beam 双悬臂托梁，99
Double lap sandwich truss 双folds夹层桁架，101
Dovetail:
　housing 燕尾榫暗榫，134
　joint 燕尾榫暗榫，106
　lap 燕尾榫明榫，107
Dowel 榫钉，79，87，107
Dowelled mitred joint 斜接暗销接头，135
Downpipe 落水管，25，125，171
Downstand 肋形，75
D.P.C(damp proof course) 防潮层，80，132
D.P.M(damp proof membrane) 防水膜，136
Drainage(land) 陆地灌溉，190
Drainage 排水，25，169

Drainage underground, 170
　(fittings——UPVC, cast iron, vitrified clay, spigot and socket, polypropylene) 地下排水（管件——硬聚氯乙烯管、铸铁、陶土管、套管与插口、聚丙烯）
Drainpipe 排水管，171
Draughting machine 制图机，9
Drawbridge 吊桥，20
Drawing board(size,types) 画图板（尺寸、类型），8，9
　equipment 作图工具，9
　instruments 作图器具，10
　pen 针管笔，10
　practice 作图实践，7
　representation 图纸表达，5
Drill(rawtool power) 冲击钻、电动钻，70
Drive cable 绳双动，9
Druidical style 督伊德教的风格，21
Drum(mixing) 搅拌轮，66
Dry lining 干砌砌，162
Duct wiring 导管线，178
Dumper(highway) 公路翻斗车，63，66

Earthquake 地震，201-203
Earth works(graphic presentation) 土方工程（图形表述），186
Easement 地役权，41
Eaves 屋檐，97
Eaves gutter 檐口天沟，125
Efflorescence 盐析，85
Egg and dart 卵与箭形装饰，161
Electricity accessories 电力附件，177
　fitting 电力装置，180
　light bulbs 电力灯泡，181
　supply and wiring 电力供应与布线，177
　track wiring 电力配线路线，178
Elevation 立面，5
Elliptic 椭圆，6
Embattled 布阵式，19
Engaged(columns) 附墙圆柱，21
English bond(brick) 英国式砌合砖，82
English land law 英国土地法，41
Energy efficient design 能量效率设计，199-200
Enrichment (classical) 古典增添装饰，18
Entablature 柱上楣构，16
Entrance 入口，33，34，35
Equipment(site) 工地装置，62

Equity(shared) 共享股本,50
Escutcheon 孔盖,149
Escalator(types) 自动扶梯(类型),128
Espagnolette bolt 长插销,149
Estimate(approximate,comparative) 估算(近似的、相比较的),51
Estimating 评估,51
Excavation plant 挖掘机,63
Excavator 挖掘机,63
Expansion tank 膨胀水箱,183
Express grant 表达授予,41
External works 户外作业,186
Extrados 拱背线,83
Extreme weather 极端天气,204-206

Faade(stone) 石质立面,87
Face(brick) 墙前面,83
Facing stone 贴面石,87
Facings 装饰面材,90
Farm building(traditional) 传统农场建筑,32
Farm house 农舍 31
Farm yard 农家宅院,31
fascia(board) 封檐板,97
Feasibility 可行性,52
Feathered end 薄端末梢,117
Fees: 费用:
 letting 租金费,49
 professional 专业人员费用,49
 sale 销售费,49
Felt(upstand) 油毡泛水,111,112
Fencing types 围栏类型,188
 (bar (vertical/horizontal),cleftchestnut,closeboarded, chainlink, palisade, woven wire/wood 木条(垂直/水平),带裂缝栗木,封闭木板,链接,编织状,钢丝网,编织状木)
Fender 壁炉挡板,131
Fibre 纤维,165
Field house 储藏室,32
Final account 竣工结算,51,53
Final certificate 决算书,53
Financial appraisal 财务评估,52
Financial aspects 财政,49
Finial 顶尖,23
Finger joint 指接节点,106
Finishes: 装修
 plasterwork 抹灰作业,158
 timber 木成品,162

Fire back 壁炉背墙,96,130
Fire bed 火床,183
Fire dogs 壁炉柴架,131
Fireplace accessories 壁炉附件,130
Fireplace recess 壁炉凹穴,133
Fireplace 壁炉,133
Fireproofing 防火,154
Fire surround 壁炉周边,96,130
FFL(finished floor level) 竣工楼板水平面,4
Fixings(metal) 金属附件,152,153
Flanking tower 侧塔,20
Flashing(metal cover) 金属盖防雨板,111,121,123
Flat arch 平拱,89
Flat roof 平屋顶,98
Flat roofs 平屋顶,111
 concrete 混凝土,111
 green 绿化,111
 metal deck 金属底模板,111
 on timber deck 木底模板上,111
 refurbished 改建的,111
Flaunching 烟囱顶泛水,121
Flemish bond 荷兰式砌合,82
Flex(two/three core) 皮线(双芯/三芯),177
Flex(lampholder) 皮线(灯座),179
Flexible curve 挠性曲线板,10
Flint, (split, knapped) 分割硬质石、碎硬质石,88
Float(metal, wood) 抹子(金属、木质),158
Floating mark 浮动测标,58
Floods 洪水,212
Floodlight projector 探照灯,179
Floor boards 地板板材,132,133
Floor drain 地面排水,169
Floor board sawing 地板的锯切,163
Floors: 楼盖
 ground, 首层地面,25
 RC/concrete 钢筋混凝土/混凝土,136,137
 raised 高架,140
 timber 木,132,133,134
 upper 上部,25
Flooring (mitred herringbone parquet wood strip) 地板(斜接人字形、拼花、木条),96
Flue 烟道,130
Flue terminal 烟道口,183
Fluorescent(tube, compact bulb) 荧光灯管、节能灯,181

Flush joint　平灰缝，81
Flying buttress　拱式扶垛，23
Flying shore　横撑，79
Footing(concrete)　混凝土墙脚，73
　(brick)　砖墙脚，80
Foundation　基础，26，67
Foundation:　基础：
　continuous　连续，75
　isolated pad　独立，74
　raft　筏板，75
　stepped　阶梯式，74
　strip　条形，73
　trench fill　坑槽填充，73
　wall pier　墙墩，73
Frame and panel(partition system)　框架墙板隔墙系统，95
Franklin point　富兰克林点，124
French curve　曲线尺（板），10
French windows　法式窗，落地长窗，25
Frieze　雕带，16，87，96
Front door　前门，25
Frog　砖面凹槽，84
Functional(tolerance, space)　功能（公差、空白），4
Funder(development)　投资者（开发），50
Funding methods　基金方法，50
Furring channels　槽形副龙骨，162
Furrowed finish　开槽的装饰，86

Gable　尖顶屋两端的山形墙，25
Gabled(barn, porch)　有山形墙的（仓房、门廊），31
Gantry mast　门式桅杆，62
Gargoyle　滴水嘴，23
Gaskets　嵌缝物，91，168
Gasket fir cone　杉木锥形嵌缝物，92
　compression　压缩嵌缝物，92
Gate house　传达室，34
Gate(postern)　侧门，20
Gateway　大门口，34
Gather　聚集，130
Gauge(marking)　划线规尺，69
Gauge(tiling)　瓦行距，114
Gazebo　凉亭，191
G.D.V.(gross development value)　总开发价值，49
G.E.A.(gross external area)　总建筑面积，49
Generator set　发电机组，67

Georgian wire glass　乔治夹丝玻璃，167
G.I.A(gross internal area)　总使用面积，49
Gland(adjusting screw)　压盖调节螺丝，184
Glass (bullet proof, flat, float, laminated, patterned, plate, rough cast, solar control, Sound control,toughened,wired)　玻璃（防弹、平板、浮法、夹层、压花、平板、辊花、日光控制、声音控制、钢化、夹丝），167
Glazing　玻璃窗，168
Glazing panel　玻璃板，91
Glazing systems(bead, gasket, patent)　玻璃安装系统（压条、嵌缝物、无油灰缝），168
Global warming and greenhouse effect　全球变暖与温室效应，195
Going　踏步宽，127
Golden number　黄金数字，3
Golden section　黄金断面，3
Gooseneck　鹅颈管，171
Gothic arch　哥特式拱，24
Gothic cathedral　哥特式主教座堂，24
Grading　颗粒级配，190
Graduation markings(metric, imperial)　刻度标号（公制、英制），7
Granary　谷仓，32
Granules(ceramic)　陶瓷颗粒，117
Grant(express, implied)　授予（表达、暗指），41
Grating　格栅，125
Gravel(path)　砾石小径，186
Greenhouse　温室，191
Greenhouse staging　温室台架，191
Green house gases　温室气体，195
Green roof　绿化屋顶，111
Grid(structural, planned)　网格（结构、规划），4
Grid(dimensionally coordinate)　尺寸坐标格，7
Grid(setting out)　网格放线，57
Grid lines　网格线，7
Groove　凹槽，89
Ground　地面，25
Ground beam　地基梁，76
Ground floor　首层，25
Guilloche　扭索饰，161
Gulley(trapped)　装有隔气弯管的集水管，171
Gusset(plywood)　胶合节点板，101
Gutter　檐沟，25，87

Hacksaw 弓锯,70
Half bat 半砖,84
Half bonding 半砌合砖,82
Half hipped barn 半斜脊的仓房,31
Half face housing joint 半叠接明榫槽节点,107
Half grid ceiling 半格栅顶棚,129
Half-lap joint 半叠接,106
Hall house(timber framed) 木框架会堂房,22
Handrail 扶手,127
Hanger:
  joist 梁托锚件,132
  steel 钢梁托,131
  timber 木吊架,101
Hanging(of windows) 窗的悬挂,146
Haunch 砖拱背圈,83
Hayloft 干草料顶棚,31,32
Header 丁砖,84
Heading course 丁砖层,81
Heart(timber) 木材中心,163
Hearth 壁炉地面,96
Hearth(front, back) 壁炉前的地板、壁炉后的地板,130
Heat exchange 热交换,197
Heat exchange pipe 热交换管,183
Heat radiation trapped 捕集的热辐射,195
Heating system 加热系统,182
Herringbone strutting 十字交叉撑,132
Herringbone mitred timber flooring 斜接人字形木地板,96
High rise building 高层建筑,30
Highway(central line) 高速公路中心线,41
Hinges (T-hinge, hook and band, butt, helical spring, loose pin, lift-off, cranked, parliament,offset) 铰链(T形、束带式、对接、双向弹簧、抽心、活脱、弯曲、H形、长翼),137,151
Hinged partition 合页隔墙,95
Hip 屋脊,113
Hip rafter 脊椽,97
Hipped(barn, porch, canopy) 斜脊的(仓房、门廊、顶棚),31
Hipped roof 四坡屋顶,98,113
Hob grate 炉盘炉排,130
Hog back ridge tile 上拱脊瓦,114
Hoist(block, rope) 起重滑车、起升钢丝绳,62
Hollow brick 空心砖,84
Honeycomb core 蜂窝式夹心墙板,95

Hook(slate fixing) 石板固定钩,113
Hopper 加料斗,183
Hopper head 水斗,125
House:
  detached 独立式住宅,30
  (field) 储藏室,32
  mass produce 大规模建造住宅,26
  stately 富丽堂皇的住宅,29
  traditional 传统房屋,25
Hot water(supply, cylinder) 热水(供应、圆筒),171
Hot water circulation principle 热水循环原理,204
Human dimensions(Le Corbusier) 人体尺寸(勒·柯布西耶),3
Hurricane development 飓风的发展,204
Hybrid power system 混合物动力系统,199
Hydraulic(lift) 电梯液压,129
Hyperbolic 双曲线,6

Imbrex 筒瓦,115
Immersion heater 浸没式加热器,175
Immersion heater circuit 浸没式加热器线路,176
Imperial(scale) 英制比例,8
Impost 拱托端,83
Indented 锯齿状,19
Indicator bolt 显示门闩,149
Infil panel 内镶板,91
Inlet 入口,183
Inspection chamber(drainage) 排水检查井,170
Instructions(architect's) 建筑师的指示,53
Integral boss connector 整体式主连接件,170
Integrated ceiling 合成顶棚,140
Interceptor 截流器,169
Interim certificates 临时执照,53
Interior 室内,75
Interstitial condensation 缝隙凝结,85
Ionic order 爱奥尼柱式,17
Ionizing point 电离辐射点,124
Ironmongery 建筑五金,148-151
  hinges 铰链,147
Ironmongery (continued) 建筑五金
  knobs and latches 把手与门锁,148
  locks 锁,148
Irregularity(notice of) 违章通知,46
Irrigation system 灌溉系统,187

Isometric projection 等角投影, 5

Jack(stabilising) 稳定千斤顶, 62
Jamb 门窗框边框, 96, 110
Jetty 外挑楼层, 22
Joggled flat arch 啮合平拱, 89
Joints: 灰缝
  bed, vertical, struck, recessed, projecting, flush, keyed, tuck, rusticated, vee, channeled 水平、竖直、刮平的、隐藏的、突出的、齐平的、凹口的、凸缝、凸形、V形、槽形, 81
  carpentry 木工, 106
  cross timber 交叉木, 107
  joggle 榫接, 87
  panel,angle,lapped and tongued 板、角接、搭接与楔形榫槽, 135
  stepped, angle, open drained baffle, sealant filled 阶梯形、角接、明排水挡板、密封填充, 92
  traditional scarf 传统嵌接, 108, 109
J.C.T.(joint contracts tribunal) forms of contract 英国建设工程合同、合同格式, 54
Joist 搁栅、托梁, 96, 132, 133
Joist hanger 搁栅锚件, 153

Keep 防守, 20
Kerb(built-up timber) 组合的木边栏, 111
Kerb(earthworks——precast, timber) 侧石（土方工程——预制、木制), 186
Keyed joint 凹圆弧缝, 81
Keystone(crown) 顶部拱顶石, 24, 89
King closer 去角砖, 84
King post 桁架中柱, 103
King post roof 桁架中柱屋顶, 99, 104, 105
Knapped flint 碎燧石, 88
Knob(door) 门把手, 144
  (ironmongery) 建筑五金, 148, 149
Knuckle joint fitting 肘形连接管件, 170

Lacing courses 拉结层, 88
Lags 木材裂缝, 163
Lamp holder(pendant,cover,flex) 悬垂灯座、灯座盖、灯座взы线, 180
Lancet 尖顶窗, 24
Land chain 测链, 7
Land drainage 陆地灌溉, 190

Land law 土地法, 41
Lap 搭接部分, 81, 82, 113, 117
Lath 拉网、板条, 154, 159
Lathing 板条, 97
Latches:
  cylinder, rim 圆筒式弹簧碰锁, 148
  lever 杠杆碰锁, 149
Lattice frame 格式框架, 62
Lead(flat, round) 铅皮、圆铅皮, 166
Leadburned saddle 烧制铅皮的鞍形座板, 122
Lead cutting knife 铅铲刀, 166
Lead sheet 铅皮, 122
Leaf and dart 叶瓣与箭头花纹装饰线脚, 161
Lean-to roof 单坡屋顶, 98
Lean-to half truss 斜半屋架, 103
Lease back guarantee 售后租回担保, 50
Le Corbusier 勒·柯布西耶, 3
Ledger(scaffold) 横木, 64, 65
Levelling staff 水准标尺, 59
Lever latch 杠杆碰锁, 149
Lever springs 杠杆弹簧, 150
Lierne 枝肋, 24
Light fitting 灯具, 180
Lighting(outdoor) 户外照明, 179
Light track 照明路线, 177
Lightening conductor 避雷装置, 124
Line(centre,break,section) 中心线、打断线、剖面线 8
L.V.D.T(linear variable differential transducers) 线性差动传感器, 61
Lines(axial) 轴线, 4
Lime(in lime and cement stucco) 石灰与水泥装饰抹灰, 160
Lintels(steel) 钢过梁, 155
Loan(repay, service) 偿还或服务贷款, 50
Local authority 地方当局, 43, 45
  (officers) 地方官员, 45
Lock(spring loaded, blocking pins) 受载弹簧闭塞销锁, 150
  (horizontal mortice) 水平榫接锁, 148
Loft 阁楼, 25
Log 圆木, 163
Loggia 凉廊, 36
Long term interest 长期利益, 50
Loop(carpet——plain, cut, sculptured) 线圈（地毯——平针、割断起毛、凹凸花纹的), 165
Loop hole 透光孔, 20

Loose box 放饲厩, 32
Lorry 卡车, 62
Lozenge 菱形, 19
Luffing(arms, rope) 俯仰式臂、变幅钢丝绳, 62
Lynch gate 教堂墓地中有顶的大门, 36

Machicolations 堞眼, 20
Magnetic field 磁力领域, 60
Magnetometry 磁力测定, 60
Mains(supply water) 给水干管, 171
Maintenance manual 维护手册, 53
Mallet 木槌, 69
Manhole(brick, cover) 砖砌人孔、人孔盖板, 171
Mansarded roof 折线形屋顶, 98
Mantle 壁炉饰面, 130
Marking gauge 划线规尺, 69
Masking tape 封口胶纸, 10
Mass movement 块体运动
  Creep 蠕变, 208
  Effect on buildings 对建筑物的影响, 211
  Fall 坠落, 209
  Flow 流动, 209
  Heave effect 隆起效应, 211
  Slide 滑动, 208
  Structural damage 结构破坏, 211
  Topple 倾覆, 210
Materials efficiency—waste management 材料效能——废弃物管理, 198
Measuring tape 卷尺, 59, 69
Mediaeval castle 中世纪城堡, 20
Mediaeval ornament 中世纪装饰, 19
Meetings(progress) 进度会议, 53
Merlon 城齿, 20
Mesh reinforcement 钢筋网, 137
Mesh(expanded steel) 拉展金属网, 154
Metal:
  components 金属部件, 154, 155
  fixings 金属设备, 153
  lath 金属拉网, 154
  ties 金属系材, 152
Meter(cover) 面层测厚仪, 60
Meter(transducer) 传感器, 61
Metope 柱间壁, 16
Metric scale 公制比例, 8
Microwave analysis 微波分析, 60
Mitred closer 切角镶边砖, 84
Mitred slates 斜接石板, 113

Modillion and patera 飞檐托与圆盘花饰, 161
Module 模数, 3
Module(basic) 基本模数, 7
Modulor 模数系统, 3
Monopitched roof 单坡屋顶, 98
Mortice and tenon joint 镶榫接头, 104
  (twin slot, open, mitred) 双槽凹凸双榫、开口凹凸单榫、斜îqİ凹凸单榫, 107
Motor pump room 马达泵房间, 129
Mouldings(timber) 木模塑, 110
Mulch(peat) 泥炭护根, 189
Mullion 直梃, 91
Multiple ownership 多种所有权, 50
Multi point lock 多点锁, 150
Muntin 门中梃, 142

Nails 钉子, 156
Nave 中殿, 23, 24
N.I.A(net internal area) 净使用面积, 49
Newel 楼梯扶手转角柱, 127
Nib(tile) 瓦的凸棱, 114
Nondestructive survey methods 非破坏性测量方法, 60, 61
Non-metropolitan(county council) 非大都市, 43
Nosing 楼梯踏步小突沿, 127
Notice of irregularity 违章通知, 46
Nozzle piece 喷嘴件, 125

Oblique projection 斜轴投影, 5
Offset bend 迂回管弯头, 125
Ogee S形曲线, 110
Ogive(arch) 尖顶拱, 24
Opening:
  roof 屋顶开口, 118
  walls 墙洞, 142
Oriel window 凸肚窗, 37
Ornament(classical) 古典装饰, 18
Ornament(mediaeval) 中世纪装饰, 19
Orthographic projection 正交投影, 5
OS(ordinance survey) 法定测量基准点, 57
Outdoor lighting(fitting) 户外照明用设备, 179
Outlet(rainwater) 雨水出口, 125
Outsider/inside 外部/内部, 36
Overage 超额, 49
Overdoor panel 门头镶板, 96

Overflow outlet 满溢出口, 170, 173
Owner occupier 自住业主, 50
Oxyacetylene set 氧乙炔气装置, 68

Pad foundation 独立基础, 68
Padlock 挂锁, 148
Pan(lavatory) 冲洗式坐式便桶, 173
Panel: 板：
　(glazing infil) （玻璃板、内镶板）, 91
　(joint) (板接头), 135
　(moulding) (板造型), 110
　(overdoor) (门头镶板), 96
Panelled door 镶板门, 144
Panelling 镶板, 164
Pantile 平瓦, 112, 115
Parabolic 抛物线, 6
Parallelepiped 平行六面体, 6
Parallel motion 平行运动, 9
Parish church 教区教堂, 23
Parquet floor 拼花地板, 96
Partitions: 隔墙
　demountable, sliding/folding hinged, frame and panel 可拆换的, 滑动折叠合页, 框架墙板, 95
　Internal non-load bearing, timber/metal stud 内部非承重木/金属立柱, 94
　Laminated, lightweight 叠层, 轻质, 93
Party fence 共用篱笆, 41
Party structure 共用构筑物, 46
Party wall 界墙, 41
Patch of sky 一方天空, 42
Patent scaffolding 专利脚手架, 64
Paving:
　concrete 混凝土铺砌面, 186
　pattern 铺面样式, 83
Pedestal 基座, 16
Pediment 三角形楣饰, 16, 144
Peg 测桩, 57, 190
Pendant(lamp holder) 悬垂灯座, 180
Permitted development 已许可的开发, 43
Perpends 竖向灰缝, 81
Perspective 透视图, 5
Photogrammetry 照相测量法, 58
Photographic measurement 照相测量 58
Pick 风镐, 67
Picked panel finish 点饰面板装饰, 86
Picture rail 挂镜线, 96, 110

Pier 墩、支柱, 73, 83
Pile(carpet)types of construction 地毯绒毛构造类型, 165
Piled foundations 桩基础, 76
Piles(shortbored, percussion, flush bored) 短钻孔灌注桩, 冲击钻孔灌注, 冲孔钻孔灌注, 76
Pillar 柱, 24
Pincers 钳子, 69
Pinfitting(electrical) 电工锁紧销, 181
Pinning(dry) 干挖掘支撑, 79
Pipe(flush) 冲洗管, 173
Pipes:
　expansion 补偿管, 174
　rainwater 雨水管, 125
　soil 污水管, 171, 173
Pipework(sanitary) 清洁的管道作业, 172
Pitch 高跨比, 98
Pitched roof(construction) 斜屋顶施工, 97
Pitched roof(types) 斜屋顶类型, 98
Pitching hole(barn) 带俯仰洞口的仓房, 31
Pivot 轴, 146
Plan 平面, 5
Planning:
　approvals 规划批准, 44
　control 计划控制, 43
　permission 规划许可, 43, 44
Plane 木工刨, 69
Plant 施工设备, 62, 63
Plant care 植物养护, 191
Plaster(levelling, scoring) 抹灰 （水准测量、划痕), 158
Plasterboard 石膏板, 93, 94, 138, 139, 162
Plaster stop 抹灰挡板, 154
Plasterwork(tools) 抹灰泥作业工具, 158
Plate(wall, head, sole, stay) 托梁垫板、顶板、底板、垫板, 65, 79, 105
Plate(pin, top, bearing, cover) 钉板、顶板、支承垫板、覆盖板, 144
Plate(switch) 开关面板, 177
Pliers 钳子, 69
Plinth 基脚, 16, 83, 89, 110
Plot ratio 容积率, 49
Plug:
　electric, safety 电路插头、安全插头, 177, 178
　partition 隔墙木楔, 92

Plugs (fixings), nylon, frame anchor, fibre, nailable 固定用栓塞、尼龙栓塞、框架锚、纤维栓塞、可钉栓塞, 157
Plunger, (disc, hook) 活塞钩、活塞圆盘, 173
Plungers 插塞, 180
Pointed arch 尖顶拱, 24
Pointed style 尖角的风格, 21
Pointing(tuck) 嵌凸缝, 81
Poker 拨火棍, 131
Poles(ranging) 测杆, 59
Porch 门廊, 24, 25, 34
Porch and chamber 门廊与会议厅, 23
Portal 大门入口, 9
Porte-cochere 车辆门道, 35
Portico 柱廊, 35
Post 支柱, 102, 103
Postern gate 暗道门, 20
Pot hook 锅钩, 131
Power socket 动力插座, 177
Power tools 动力工具, 67
Power track 动力路线, 178
Practical completion 操作完成, 53
Precast cladding(panel fixing) 为预制覆面镶板固定, 155
Prescription 剥夺公权, 41
Pressed brick 压制砖, 84
Priority yield 优先收益, 50
Prism 棱柱体, 6
Profit erosion 利润侵蚀, 50
Project:
  design 项目设计, 52
  management 工程管理, 50
Projecting joint 突出缝, 81
Projection (axonometric, isometric, oblique, orthographic) 投影（不等角、等轴测、斜轴、正交）, 5
Projector(flooding) 探照灯, 179
Proportions(module) 模数比例, 3
Proprietary cubicles 专用小室, 95
Protractor 量角器, 9
P-trap P形存水管, 174
Pulley wheel 滑轮, 9
Punch 冲孔器, 70
Punched finish 冲孔的装饰, 86
Purlin 檩条, 97, 101, 102
Pyramid(truncated) 斜截棱锥, 6

Quadrant moulding 扇形板造型, 110
Quadripartite(vault) 分成四组的拱顶, 23
Quality indicator 质量指标, 51
Quantities(approximate) 近似工程量, 51
Quarry:
  bottom, limestone 矿层底部、石灰石矿层, 86
  glass 方形玻璃片, 166
Quarter bonding 1/4 砌合, 82
Quarter circle 四分之一圆, 19
Queen closer 纵半砖, 81, 82, 84
Quoin:
  brick 砖隅石, 88
  headers 角砖, 86
  picked panel 点饰面板隅石, 80

Rabbet 凹凸榫接, 161
Radiators (panel, column, oil filled, storage, heater) 板式散热器、柱式散热器、电热油汀、蓄热加热器, 182, 184
Radiography 射线照相术, 60
Radius 半径, 83
Raft foundation 筏基, 75
Rafter 上弦木, 97, 102, 103, 104, 105
Rail (door, lock, frieze) 门横档、装锁横档、雕刻条板下横档, 144
Rail(picture) 挂镜线, 96
Raised floors 高架地板, 141
Raking shore 斜撑, 79
Ram 活塞, 62, 63
Rammer 夯具, 67
Random coursed 随机成层的, 88
Rat-trap bond 空斗墙砌合, 82
Ready-mix concrete(truck) 预拌混凝土车, 66
Rainwater pipes 雨水管, 125
Rebate(glazing) 门窗框槽玻璃窗, 168
Recessed joint 方槽凹缝, 81
Reflector bulb 反射灯泡, 181
Reinforcement(bar, mesh) 钢筋、钢筋网, 137
Relief, (decorative) 装饰浮雕, 20
Rent 租金, 50
Rental income 租金收入, 50
Rental growth 租金上涨, 49
Representation(drawing) 作图表现, 8
Reservation 保留, 41
Retaining walls, (small, medium, basement) 挡土墙（小型的、中等的、地下室的）, 77

Retention 保留金,53
Reticulated finish 网状的装饰,86
Reverse bond 反向砌缝,82
Rib: 肋
  lightweight precast 轻质预制肋,92
Rib:(continued) 肋(继续)
  longitude/transverse 纵肋、横肋,137
  ridge 脊肋,23
  tierceron 居间肋,24
Ridge(beam) 脊梁,97,98,102,103
Rights of:way, light, support 路权、灯光权、赞助权,41
Rings(arch) 拱环,83
Rise 拱高,83
Riser 梯级竖板,127
Rivet 铆钉,157
Roach 岩石、罗奇层,86
Rod 拉杆,101
Rodding eye 通管孔,169
Roll(lead) 铅卷筒,122
Roman tiles 罗马式瓦,112,115
Roof:
  cover, structure 屋上盖、屋顶结构,25
  covering type 屋面覆盖层类型,
  pitched construction 斜屋顶施工,97
  run, span, rise 屋顶宽度之半、跨度、垂直高度,98
Root, system, bare, balled 根系、裸露的根、再包土的根,189
Rose window 圆花窗,24
Rubble 毛石:
  backing 背衬,87
  bottom 底部,86
  random, coursed, square 随机、成层、方块,88
  work 砌体,88
Rusticated joints 凸形勾缝,81

Saddle bar 撑棍,166
Safety plug 安全插头,178
Sanctuary 圣堂,24
Sandwich joint 夹板节点,104
Sanitary pipework 卫生管道作业,172
Sap(log) 原木的树液,163
Sash:
  movable 活动式窗扇,120
  window 格窗,146

Saw 锯子,70
Sawing(rift, tangential) 四开锯木、切向锯开,163
Scaffolding:
  board 脚手架板,65
  independent 独立脚手架,65
  patent 专利脚手架,64
  tower 脚手架塔,64
Scale(metric, imperial) 公制比例、英制比例,8
Scale rule 比例尺,9
Scalpel 手术刀,10
Scarf joints(traditional timber) 传统木嵌接节点,108,109
Scissor(beam roof) 剪式梁屋顶,99
Scoring nails 刻痕钉,158
Scotia 凹痕线,18
Scratcher 划痕器,158
Screening(landscape) 景观屏蔽,186
Screw cap 螺帽,181
Screwdriver 螺丝刀,70
Screws (thread cutting, self drilling, head types) 自切螺钉、自钻螺钉、螺钉头部类型,156,157
Scull cap 圆盖层,86
Sea breeze principle 海风原理,204
Seasoning(timber) 木材干燥法,163
Semidetached(building) 半独立式建筑,27
Septic tank 化粪池,169
Set-square(adjustable) 可调节直角尺,9
Setting out grid 网格放线,57
Shaft(column) 柱身,16
Shallow void floor 浅空隙楼板,141
Shapes 形状,6
Shared equity 共享股本,50
Shed(open fronted) 前开式棚,32
Shelter shed 掩蔽棚,32
Shelter(wind) 防风罩,187
Shingle 砾石,86
Shingles(roof covering) 屋面板(屋顶覆盖物)
  felt, edge grained, flat sawn 油毡屋面板、径锯屋面板、顺锯屋面板,117
  traditional timber 传统木屋面板,117
Shoe 落水斜口、门脚,125,145
Shoring 支撑,79
Shoulder 榫肩,106
Side hung window 侧旋窗,145

Sill 门槛, 142, 143
Simplified planning zone 简化的规划区域, 43
Siphonage 虹吸作用, 172
Site(building) 建筑工地, 55
Site visits 工地参观, 53
Skirt 裙板, 180
Skirting 踢脚板, 93, 96, 132
Skirting types 踢脚板类型, 110
Sky, angle, factor 天空角度、天空因素, 42
Skylights(proprietary) 专用天窗, 120
Slat 石板, 86
Slate 石板瓦, 112, 113
Sleeper wall 地垄墙, 132
Sleeve 套筒, 125
Slewing rig 回转器, 62
Sliding/folding partition 滑动/折叠隔墙, 95
Sliding sash 推拉窗, 145
Smoke detector 感烟探测, 140
Snib 插销, 149
Soakaway 污水渗透坑, 190
Socket:
  circuit, double 插座电路、双插座, 176
  cover 插口罩, 181
  outlet 插座出口, 178
Soffit 拱圈内面、挑檐底面, 83, 97
Soffit cleat 下表面加强角片, 139
Soil fitting(UPVC) 地下连接件（硬聚氯乙烯管）, 170
Soil and vent stack 地下排泄竖管, 172
Solar power 太阳能, 199
Solar radiation 太阳辐射, 195
Soldering(iron) 焊铁, 68
Sole plate anchor 底板锚件, 153
Space for services 设备空间, 140
Span 跨, 83
Spandril 拱肩墙, 83
Spanish tiling 西班牙式瓦, 115
Spanner, (box) 套筒扳手, 69
Spar 椽子, 115, 116
Sphere(segment) 球体分割, 6
Spigot and socket soil fitting 套管与插口地下接头, 170
Spike 尖锋, 97
Spindle: 轴杆
  locks 锁, 145
  plumbing 自来水管道工程, 173
  Wilk's, Duce's, patented, slotted, floating,

Pitt's candle 转轴：威尔克、达斯、专利的、开槽、浮动、皮特式烛形物, 148
Spire 尖塔, 23, 24
Spirit level 气泡水准仪, 70
Splay:
  door 门斜削面, 96
  header, stretcher 斜削丁砖、斜削顺砖, 84
Splayed and rounded skirting 抹角的与圆的踢脚板, 110
Splayed roof 八字形屋顶, 98
Splayed shouldered lap joint 八字榫肩明榫, 107
Split end tie 裂端系杆, 152
Split hazel 分割的榛木, 116
Spot(low wattage) 低瓦数地点, 178
Spotlight 聚光灯, 178
Springing line 起拱线, 83
Sprinkler 喷洒器, 146
Sprocketed roof 接椽式屋顶, 98
Square(adjustable) 可调尺, 69
Square housed joint 方形暗榫槽节点, 107
Staircases: 楼梯
  special (samba, cat ladder, retractable, loft) 特殊的（交错式、爬梯、可缩式阁楼爬梯）, 128
  timber 木的, 127
  types (straight flight, dog-leg, open well, spiral, bifurcated, quarter turn) 类型（直跑楼梯、层间双折楼梯、露明梯井、螺旋楼梯、分叉楼梯、直角转弯楼梯）, 126
Stable 畜舍, 31, 32
Stable, (lofted) 带阁楼式马厩, 31
Stack(chimney) 烟囱, 25
Staff 标尺, 58, 59
Stained glass 彩色玻璃, 166
Stake 木桩, 189
Stanchion 支柱, 70
Stand(drawing board) 画图板支座, 9
Stanley knife 斯坦利刀, 70
Star 星形, 19
Starter 启辉器, 181
Statutory(undertakers) 法定办者, 46
Stave 板条, 159
Stave socket 板条承接头, 159
Steel:
  beams 钢梁, 78
  column foundation 柱下型钢基础, 78
  grillage foundation 型钢格排基础, 78

wedge 钢楔底板,78
Stencil 誊写版,10
Stepped foundation 阶梯式基础,74
Steps 台阶,25
Stereo plotter 立体绘图仪,58
Stile(door) 门梃,144
Stone 石材,86
Stone facing 贴面石,87
Stone surface finish 石材表面装饰,86
Stone tiles 石材瓦,112
Stone walling 石材墙体,88
Stool 窗台线端部找平座,89
Stop cock 开关,171
Stop end 末端,125
Stop valve 停止阀,171
Storage heater 贮存加热器,175
Storage tank 贮藏水箱,171,174,183,
Storm clouds 暴风云,204
Straightening rule 划线尺,158
Strainer flow control 滤网流动控制,187
Stretcher 顺砖,84
String course 带饰,87,89
Strip foundation 条形基础,73
Struck joint 下斜缝,81
Structural grid 结构网格,4
Strut 支撑,97,101,102,105
Strutting(herringbone) 十字交叉撑,132
Stucco 装饰抹灰,160
Stud partition 隔墙立柱,93
Stump(tree) 树墩,163
Styles compared 风格比较,21
Stylobate 柱列台座,16
Sub-structure 地下结构,25
Sun light 阳光,42
Sun light indicator 阳光指示器,42
Sun light spacing criteria 阳光空间标准,42
Sun shades 遮阳罩,140
Survey(accurate) 精确测量,58
Sustainable 'Green' buildings 可持续绿色建筑,196
Swan neck 鹅颈形弯头,125
Switch(electric——dimmer, one-way) 电的调光器开关、单向开关,177
Swivel coupler 旋转联结器,64
Swivel handle 活动头手柄,7

Tally 记录,7

Tamping board(vibrator) 强夯板式振动器,66
Tank:
  cold water 冷水箱,174
  expansion 膨胀水箱,183
Tape measure 卷尺,69
Tee square 丁字尺,9
Tegula 沟瓦,115
Template 模板,10
Temple(classical) 古典庙宇,16
Temporary bench mark 临时基准点,58
Tender documents 投标文件,52
Tender stage 投标阶段,51,53
Tenon: 凸榫,134
  bare faced 半肩榫,107
  double shouldered 双肩榫,107
  oblique 斜榫,106
Tenement(Scottish) 苏格兰式分租合住的经济公寓,28
Terraced building 联立式建筑,27
Tetrahedron 四面体,6
Thatch 茅草,112,116
Theodolite 经纬仪,57,59
Third bonding 1/3 砌合,82
Three cell(plan type) 三单元类型平面,23
Threshing barn 脱谷板棚,32
Throating 滴水槽凹口,89
Tie:
  beam 系梁,97,101,102,103,104,105
  diagonal 斜拉杆,97
  metal——stainless steel, bow 金属——不锈钢弓形材料,152
Tierceron rib 居间肋,24
Tiling(roof plain) 屋顶平瓦,114
Timber:
  batten 木板条,93
  conversion 大木料锯小,163
  deadmen 木地锚,189
  door frame 木门框,142
  floor construction 木楼面结构,132
  framed buildings 木结构框架建筑,22
  roof construction 木屋顶建设,97
  seasoning 木干燥法,163
  stud 木立柱,93
Tolerance(functional) 功能容限,4
T&G(tongue and groove)joint 槽舌榫接头,135
Tools 工具,69,70

Tools(plasterwork)　抹灰泥作业工具, 158
Tooth　齿状, 19
Toothing　留马牙槎, 81
Top cap,　石层之上的土层, 86
　rubble,　顶层碎石, 86
　soil　表层土, 86, 190
Torus　圆盘线脚, 18, 110
Tower(flanking, corner, watch)　侧面塔、角塔、瞭望塔, 20
Tower(church)　教堂塔, 23
Track(wiring, power)　布线路线、动力路线, 178
Traditional:
　scarf joints　传统嵌接节点, 108, 109
　timber roofs　传统木屋顶, 99
Transept　教堂的十字形翼部, 24
Transom　门窗横档, 91, 147
Transducer monitoring　传感器监视, 61
Trap outlets (U, bottle, P, S)　U形存水管、瓶状存水管、P形存水管出口、S形存水管出口, 172, 173, 174
Tree:
　forms　树形, 189
　guying　树的牵索调位, 189
　nursery　树苗圃, 187
Tread　踏板, 127
Trellis　格架, 191
Trench-fill foundation　坑槽填充基础, 73
Triforium　教堂拱廊, 23
Triglyph　三槽板浅槽饰, 16
Trimmed openings　饰边开洞, 133
Trimmer　托梁, 118, 120, 121, 133
Trimming of floors　木地面镶边, 134
Trivet　三脚火炉架, 131
Truncated (pyramid, prism, cylinder, cone)　斜截棱锥、斜截棱柱、截圆柱体、截头圆锥体, 6
Truss(bolt and connector)　螺栓和连接器桁架, 102
Truss(lean-to half)　斜半屋架, 103
Truss(nailed timber)　钉固木桁架, 101
Trusses　桁架, 100, 101, 102, 103, 104, 105
Truss clip　桁架夹子, 153
Truss-out scaffold　桁架外部脚手架, 65
Trussed rafter　桁架式椽条, 100
Trussed purlin roof　桁架式檩条屋顶, 101
Trysquare　直角尺, 70
Tungsten halogen bulb　钨丝卤化灯泡, 181

Turf(laying of)　草皮铺设, 190
Turnbuckle　螺丝扣, 189
Turret　塔楼, 20
Tuscan order　塔司干式柱式, 17
Tusk　齿状物, 134
Tympanum　山墙饰内的三角面, 16

U.D.P(unitary development plan)　单一的开发方案, 43
Underpinning(legs, wall)　托换基础支脚、托换基础墙, 79
Unitisation　联合经营, 50
Urban development area　城市开发区, 43
Use classes　使用级别, 43, 44

Valley　天沟, 97
Valley(rounded)　圆形天沟, 113, 123
Valve:　阀门
　control, stop, drain　控制阀、停止阀、排水阀, 183
　diaphragm　隔膜阀, 174
　Portsmouth, Croydon　朴次茅斯阀、克劳伊登阀, 174
　sequencing　程控阀, 187
　thermostatic(radiator)　恒温散热器阀, 182
　Vapour barrier　防潮层, 111
Vault:
　quadripartite　分成四组的拱顶, 24
　ribbed　带肋拱顶, 24
Vee joint　V形勾缝, 81
Verge board　山墙封檐板, 97
Vermiculated finish　虫饰状琢凿面的装饰, 86
Vertical plane　垂直面, 5
V groove　V形槽, 86
Vibrator tamping　强夯振动器, 66
View direction　视线方向, 5
Vitrified clay(drainage fittings)　陶土管排水接头, 170
Vitruvius　维特鲁威, 3
Volcano eruption　火山爆发, 213
Volumes　体积, 6
Voussoir　拱楔块, 83, 89

Wall:
　cavity　空心墙, 26, 80
　interior　内墙, 162
　plate　托梁垫板、卧梁, 65, 79, 97, 102

pier foundation 墙墩式基础, 73
retaining 挡土墙, 77
Waste systems 排废水系统, 171
Water heating (electric storage/instant, gas instant, indirect, and immersion) 电贮存水加热、电力即时加热、燃气即时加热、间接加热、浸没式加热, 175
Waters:
　leaf 水叶装饰, 161
　main 给水干管, 171
　supply(hot and cold, plumbing) 水供应(热与冷、管道), 171
　tank 水箱, 174
Warranty(collateral) 相关的保证, 54
Watch tower 瞭望塔, 20
Wattle and daub 篱笆与涂抹, 159
Wave energy 波浪能, 200
Weather bar 挡水条, 142
Weathering 泻水面, 89
Weather proofing 防风雨, 122
Weep holes 泄水孔, 80
Weights(sash) 格窗重物, 146
Welt 盖缝条, 122
Wetted strip 浸湿带, 187
Wheelbarrow 独轮手推车, 66
Whit bed 波特兰岩层, 86
Wind effect on buildings 建筑上的风效应, 205
Wind effect on roof shape 屋顶形状的风效应, 205
Winders 螺旋楼梯踏步, 126

Window board 窗台板, 147
　bay, bow 凸窗、弓形窗, 37
　casement 竖铰链窗、平开窗, 147
　cill 窗台, 89
　dormer 天窗, 118, 119
　frame 窗框, 146, 147
　French 落地长窗, 25
　leaded 铅条镶嵌玻璃窗, 166
　oriel 凸肚窗, 37
　rose 圆花窗, 24
　sash 格窗, 146
　traditional 传统窗户, 37, 146, 147
Wire-cut brick 钢丝切割砖, 84
Wire hanger 金属线吊筋, 139
Woodblock 木地板, 96
Wood strip flooring 木条地板, 96
Work(removed or hidden) 作业移除或隐藏线, 8
Working size(component) 组件的工作区尺寸, 4

X-ray source X射线源, 60

Yarn types(carpet) 地毯纱线类型, 165

Zinc:
　flashing 锌铁泛水, 123
　sheet liner 衬砌锌铁板, 123
　soaker 锌铁泛水板, 123
Zone 区域, 4

参考文献

Architect's Data, Ernst Neufert, Crosby Lockwood Staples (1970).

Architect's Legal Handbook, Anthony Speaight and Gregory Stone, Architectural Press (1998).

Building Construction Vols I,II,III and IV, W.B. McKay, Longmans (1995).

The Building Design Easy Brief, Henry Haverstock, Morgan Grampain (1987).

Building Construction Handbook, R. Chudley, Laxton's (1988).

The Care and Conservation of Georgian Houses, Architectural Press with Edinburgh New Town Conservation Committee, Paul Harris Publishing (1978). Architectural Press (1980).

Dicobat – Editions Arcature (1990).

Dictionaire – Librarie Larousse (1981).

Drawing Office Practice for British Standard 1192, Architects and Builders (1953).

Ecohouse a design guide – Sue Roaf, Architectural Press (2002).

English Historic Carpentry, Cecil A. Hewett, Phillimore (1980).

Farms in England, Peter Fowler, Royal Commission on Historic Monuments, HMSO (1983).

Handbook of Urban Landscape, Cliff Tandy, Architectural Press (1975).

History of the English House, Nathaniel Lloyd, Architectural Press (1975).

Mitchell's Building Series,

  Structure and Fabric 1, Jack Stroud Foster (1973).

  Structure and Fabric 2, Jack Stroud Foster and Raymond Harrington (1976).

  Components, Harold King (1983) Batsford Academic and Education.

Modern Practical Masonry, E.G. Warland, Sir Isaac Pitman & Sons Ltd., 2nd edn (1953).

Modulor Le Corbusier, Faber & Faber (1951).

New Metric Handbook, Edited by P. Tutt and D. Adler, Architectural Press (1979).

The Parish Churches of England, Charles Cox, B.T. Batsford (1954).

The Penguin Dictionary of Building, John S. Scott, Penguin (1982).

Repair Manual Reader's Digest (1976).

Sewage solutions – Nick Grant, Mark Moodie, Chriss Weedon – Centre for Alternative Technology Publications (2000).

Specification 1 – 6 Architectural Press (1987).

Sustainable Architecture – Brian Edwards, Architectural Press (1999).

Traditional Farm Buildings Richard Harris, Arts Council Exhibition Catalogue (1982).